RESPONSIVE LIVING

GOLDEN RAY
BOOK PUBLISHING

Responsive Living

Golden Ray Books is an imprint of Golden Ray Publishing, Inc.
7000 N. Central Expressway
Suite 501
Cheyenne, WY 82001

Printed in the United States of America
ISBN 979-8-9945080-2-2 (paperback)
Editing, Copyediting, Proofreading, Text design and Composition, Cover Design, Cover Photo by:
Golden Ray Publishing, Inc.

Printed by : Ingram Sparks
Distributed digitally by Target, Walmart, Barnes & Noble Bookstores, Amazon, Kindle, Indie Book Sellers, and Amazon Audible for Spotify and Apple.

Disclaimer

This book is intended for informational purposes only and does not constitute medical advice. The information provided is not a substitute for professional medical care. Readers should consult with a qualified healthcare provider before making any decisions related to their health or treatment. The author and publisher are not responsible for any errors or omissions in the content or for any consequences resulting from the use of this information. The author is not a medical professional, and this book is not intended to be a substitute for professional medical advice.

'Between stimulus and response there is a space. In that space is our power to choose our response. In our response lies our growth and our freedom.'

– Viktor E. Frankl

CONTENTS

Foreword

Over the course of my career in healthspan, business and performance, I've been privileged to cross paths with some of the most insightful minds in human optimisation. Few stand out as profoundly as Justin Maguire. His ability to translate complex immunological and biochemical science into practical solutions is nothing short of extraordinary.

When I first worked with Justin, it wasn't just about information, it was about transformation. I saw firsthand how his approach to addressing toxicity, hormonal imbalance, gut dysfunction, and even the mental blocks that derail so many of us, could restore clarity, energy, and resilience in ways that traditional methods often overlook. His work has helped not only me but also many of my clients, bringing resolution where there was confusion, and progress where there had been stagnation.

This book reflects that same brilliance. Justin unpacks the hidden drivers of immune dysfunction, particularly the modern epidemic of Th2 dominance, in a way that's both rigorous and accessible.

He doesn't just diagnose the problem; he maps the pathways back to balance, resilience, and vitality. What you'll find here isn't theory for theory's sake, but a framework for reclaiming health in the face of environmental stress, lifestyle pressures, and biological disruption.

What I respect most about Justin is that he bridges worlds – science and practice, research and lived experience, the complexity of human biology and then the necessary actionable steps to follow. He has a gift for connecting dots that most would never see, and an integrity that ensures his insights always remain grounded in service to others' well-being.

As you read these pages, prepare to shift your perspective on health. This isn't a quick fix or another trendy protocol; it's a road-map that can help you understand your body at a deeper level and unlock its capacity for repair, resilience, and longevity.

I'm honoured to call Justin both a colleague and a friend, and I believe the knowledge within this book will be as life changing for you as his work has been for me and those I've guided.

KRIS GETHIN
Voted World #1 Transformation Expert
Co-founder of Unmatched Supplements
Co-founder of Kris Gethin Gyms
Host of the Kris Gethin Podcast

Foreword

Chronic disease is the epidemic of the twenty-first century. The majority of suffering today does not come from acute infection or trauma, but from the silent miscommunication between our immune system and the modern world. Chronic psychological stress, environmental toxicity, disrupted sleep, industrialized food, and static physiology have created a biological climate in which the immune system becomes hypervigilant to the harmless and inattentive to the dangerous.

Justin Maguire's *Responsive Living* is a timely and important contribution to correcting this imbalance. The narrative woven through the character of James reflects the lived experience of millions whose fatigue, inflammation, allergies, cognitive dysfunction, and sensitivities have defied traditional diagnosis. Justin traces these symptoms back to their true origin: a disruption of immune balance driven by mitochondrial distress, chronic inflammatory triggers, and unresolved cellular survival responses.

This work stands apart because Justin correctly integrates the most overlooked piece of this puzzle: oral-systemic health. He dedicates a full chapter to silent dental inflammation, infectious root canals, and jawbone cavitations as persistent activators of chemokines like RANTES that drive Th2 dominance. This reflects advanced understanding that is largely missing in conventional healthcare. Justin knows this firsthand. He pursued comprehensive treatment of his own cavitations directly with me, not as a theoretical concept but as a personal cornerstone of his own healing journey.

Justin speaks with experience, precision, and a profound respect for the body's innate healing capacity. His expertise in immune modulation and clinical biochemistry has earned him the reputation of a true blood detective, and he is one of my trusted friends and closest peers in the shared mission of global health optimization.

This book provides a clear pathway to restore immune rhythm and cellular resilience. It reminds us that the immune system is not broken. It is responding appropriately to a world that has changed faster than our biology. Through the removal of hidden inflammatory triggers, support for mitochondrial efficiency, nervous system recalibration, and the reintegration of oral health as a foundational system, recovery becomes not only possible but expected.

Responsive Living empowers readers with the understanding required to reclaim their biology and return to a state of true adaptability.

DR DOMINIK NISCHWITZ
Founder of the Biodentistry Global Standard (BGS)
World-Leading Biological Dentist
Author of *It's All in Your Mouth*

Foreword

I first met Justin McGuire in 2020. What began as a professional relationship quickly became a friendship, because Justin isn't just another practitioner, he's one of the most deeply caring humans I know and I'm lucky to call him one of my closest friends, sharing the same mission to help the world be healthier and happier one person at a time. He wants the best for everyone around him, often at his own cost. Over the years, he's been a keynote speaker and panellist at the Health Optimisation Summit, and he's worked with me personally on some of my toughest health challenges of all (Thankfully!)

One moment stands out. In 2024, I suffered INTENSE gut issues for six weeks. Various experts prescribed probiotics, peptides, and other interventions that got me nowhere. I called Justin. Within minutes, he said one word: 'parasites'. He told me to get Alinia immediately. Within 24 hours my gut was back to normal. He was right, it traced back to sushi I'd eaten six weeks earlier. That single insight didn't just solve the problem, it also fuelled some of my most viral content around parasites - and his work (mind, body and spirit) fuelled a huge portion of my personal growth too. That's Justin: the 'blood detective' with the empathy of a monk and the depth of knowledge of ten incredible practitioners rolled into one.

Everyone has heard of the immune system. Few understand its branches, or why some people's immune responses become overactive and cause chronic issues. Justin breaks this down with clarity, using metaphors like TH1 artillery and TH2 soldiers that anyone can grasp. And he doesn't stop at theory. His perspective comes from thousands of hours working with hundreds of the world's most optimised humans including myself. It's a rare blend of natural health, spirituality, biohacking, and pharmaceutical science applied with systems thinking.

That's what makes *Responsive Living* so important. This isn't a book of 'do this, don't do that'. It's a framework for understanding how mitochondria, hormones, and immune pathways interlock, and how to respond when the body's signals get out of rhythm. I often wish a book like this had been available when I first got sick in 2012.

So read these pages closely. What you'll find here is not just information, it's coherence, clarity, and hope. Justin has given us a map back to the ultimate resilience and lasting lifespan.

TIM GRAY
UK's Leading Biohacker
Founder, Health Optimisation Summit (UK & USA)
400M+ annual content views
8,000+ attendees across two continents

Foreword

It has been a privilege to watch Justin develop from a promising student into a thoughtful practitioner whose understanding of cellular biology is both profound and deeply integrative. His scientific rigor is matched by a genuine commitment to translating complex biological systems into frameworks that restore health and coherence in his patients.

Responsive Living reflects this maturity of thought. It bridges the divide between molecular science and lived human experience, offering a clear and comprehensive view of how our biology mirrors the environments we create. By exploring the dynamic interplay between mitochondria, hormones, and immune pathways, the book reframes common modern conditions – fatigue, allergy, and autoimmunity – not as isolated disorders, but as adaptive responses to imbalance.

Through its scientific precision and human insight, *Responsive Living* invites clinicians and readers alike to see the immune system as a reflection of life itself: responsive, intelligent, and capable of renewal when given the right conditions. It stands as an important contribution to the evolving field of integrative and functional medicine, and a testament to Justin's dedication to advancing both the science and the art of healing.

DR AHMED HAFFEJIE
Integrative and Functional Medicine Practitioner MBBCH (Wits) ABAARM (American Academy of Anti-aging and Regenerative Medicine)

Foreword

Justin uses 'James' story', a compelling case study that represents the experiences of many individuals facing similar internal struggles, as the central thread. This approach provides a deeply human and relatable context for an otherwise intricate scientific discussion. The narrative is almost poetic in its flow, making it a surprisingly pleasant and engaging read, despite the density of the subject matter.

At its core, the book presents an excellent, comprehensive review of functional health, told primarily through the lens of immune system T-helper 2 (Th2) dominance. The author meticulously demonstrates the profound interconnectedness of systems biology. It provides a powerful refresher on how foundational systems, such as the immune system, endocrine and nervous system, are inextricably tied together. The book shines by illustrating how a perturbation in one system, such as a Th2-driven immune shift, cascades to affect virtually every other system and body function. The narrative culminates in a fascinating revelation of the internal battle of functions, detailing the clinical trajectory that ultimately leads to James's resolution and relief.

The most significant contribution of this work lies in its exposition of the therapeutic solution: the need to develop a profound capacity for interpreting and understanding the body's physiological signalling. The book provides a step-by-step resolution strategy that focuses on achieving homeostatic balance. This is done through a carefully considered and delicate combination of targeted natural and pharmacological interventions. This approach focuses on modulating core physiological dysfunctions (addressing the root causes) rather than merely palliating symptomatic expressions, effectively shifting the internal systems back toward equilibrium.

A fascinating takeaway is that the book provides such an exhaustive and integrated 'recipe' for systemic modulation. While the sheer scope of interventions would likely prove impossible to implement simultaneously in a typical clinical setting (due to time constraints, complexity or clinical impracticality), this comprehensive approach is precisely what makes the text so valuable: it gives practitioners a full arsenal of evidence-based tools to select from for their individualised patient protocols.

This text may serve as an invaluable refresher for experienced health practitioners seeking to deepen their understanding of the topic and wider chronic, multifactorial conditions. Interestingly, by correcting common misconceptions in clinical practice and connecting the dots between disparate physiological processes, the book encourages a more integrative while still scientifically grounded approach to patient care.

Beware readers: those of you without a foundational knowledge of human physiology and immunology may find the scientific detail, though well-explained, to be a steep but rewarding learning curve.

In summary, *Responsive Living*, by Justin Maguire, is a powerful work that seamlessly blends rigorous science with compassionate clinical storytelling, offering essential insights into the complex web of human health.

ANNA PAPOUTSA SHUE
Scientist, Clinical Director & Founder, Nuthrive

Acknowledgements

I wish to express my deepest gratitude to the many individuals who have guided, inspired, and supported me throughout this journey.

To Dr Dominik Nischwitz, Prof. Robert Naviaux, Dr Christopher Shade, Dr Bruce Patterson, Dr Kurt Woeller, Dr Ahmed Haffejie, Dr Rory Jean Jacques, Dr Paul Cheney, Dr William Seeds, Dr Ann Lannoye, Dr John Lieurance, and Anna Papoutsa – thank you for your pioneering work and the wisdom you have shared, which has profoundly shaped my understanding and practice.

To Tim Gray and Kris Gethin – thank you for being a source of personal inspiration, showing through action and example what dedication to health, performance, and transformation can achieve.

I am indebted to Sonja Buckland, my assistant, whose dedication and ability to keep me grounded has been invaluable.

To Lisa Ray, for making publication possible, and to Annette Greve and Isabel Von Fritschen for their thoughtful editorial support, I thank you.

A special mention to my family: my wife Albina, whose continual support sustains me; my son Logan, whose inspiration gives my work deeper purpose; and my mother Isabel, whose unrelenting faith has been my anchor.

Above all, I give thanks to Christ Consciousness – for renewal, clarity, and the shift in awareness it has brought, and to all those who have entrusted me with the opportunity to serve health – this work is for you.

JUSTIN GREGORY MAGUIRE
BSc (Hons) Nutritional Science · PGDip Functional Blood Chemistry Analysis · BTech Kinesiology & Applied Anatomy
Autonomic Wellness

Introduction

The Story of Th2 Dominance – A Modern Challenge to Immune Resilience

The human immune system stands as a testament to the brilliance of evolutionary design. It wields an extraordinary capacity to defend against a vast array of external threats, adapting seamlessly to new challenges while recalibrating to maintain equilibrium. Yet, despite this remarkable adaptability, the pressures of modern lifestyles and environmental changes have disrupted the delicate balance essential to its efficiency. At the heart of this disruption lies Th2 dominance – a skewed immune response that has become a defining feature of numerous chronic health conditions in the modern era.

This book delves into the phenomenon of Th2 dominance, unravelling its origins, the impact of static physiology during periods of neurological stress, and the critical role of immune diversity in fostering balance. It tells the story of an immune system built for dynamic adaptability yet often trapped in inefficiency, its ability to protect and heal compromised by the demands of modern living.

What is Th2 dominance?

When the immune system achieves balance, it can effectively combat extracellular threats (assisted by Th2 cells) without compromising its ability to eliminate intracellular pathogens or repair damaged tissues (assisted by Th1 cells). However, disruptions caused by chronic stress, poor nutrition, environmental toxins, or inactivity derail this recalibration. The result is Th2 dominance, marked by inefficiency, inflammation, and chronic disease.

For millennia, the immune system of our ancestors evolved to meet immediate, life-threatening challenges – bacterial infections, parasitic invasions, or injuries encountered in the natural world. Their physiology thrived in an environment that required constant movement, exposure to diverse microbes, and periodic recovery from stress. The immune system functioned in a delicate, dynamic balance, safeguarding against external threats while preserving internal harmony.

In contrast, the immune challenges we face today are fundamentally different. The threats are no longer acute or physical but rather subtle, persistent, and often self-imposed. Chronic stress, sedentary lifestyles, over-sanitised environments, processed diets, and continual exposure to environmental toxins have created conditions ripe for immune dysregulation. Within this context, Th2 dominance has silently emerged as a widespread issue, contributing to allergies, asthma, autoimmune diseases, and chronic inflammation.

One of the most profound drivers of Th2 dominance is the modern disconnect between how we experience stress and how our physiology responds to it. Unlike the acute physical challenges of the past that required bursts of energy for recovery, today's stressors – work deadlines, financial worries, or information overload – are predominantly psychological. These relentless mental stressors, unaccompanied by corresponding physical exertion to dissipate them, leave a lasting impact on the nervous and immune systems.

Under prolonged mental stress, the hypothalamic-pituitary-adrenal (HPA) axis is activated, leading to sustained cortisol secretion. Initially, this suppresses Th1 immunity, which is critical for combating intracellular pathogens and maintaining tumour surveillance. Over time, the prolonged suppression of Th1 responses strengthens Th2 dominance, skewing the immune system towards humoral activity, whereby antibodies neutralise pathogens and mark them for destruction by other immune cells. However, this shift creates a physiological state primed for allergic and inflammatory reactions but less capable of addressing systemic threats like viral infections or cancer.

Static physiology – where physical inactivity accompanies mental stress – further exacerbates this imbalance. Without the physical exertion that once accompanied stress responses, the body becomes locked in a cycle of heightened Th2 activity and diminished resilience. In this state, even the most robust immune system falters, its efficiency undermined by a lack of dynamic balance.

The immune system is designed to adapt, defend, and repair, holding the potential to protect us from virtually any threat, whether external pathogens or internal dysfunctions. However, this potential hinges on efficiency – the ability to allocate resources effectively, balancing responses without over-compensating or underperforming.

Th2 dominance disrupts this efficiency. The immune system becomes trapped in a cycle of overactivity in certain areas – producing Immunoglobulin E (IgE) antibodies, releasing histamine, and driving inflammation – while neglecting vital tasks like cellular repair, pathogen clearance, and immune regulation. This inefficiency creates fertile ground for chronic allergies, autoimmune disorders, neuro-inflammation, and even metabolic and endocrine dysfunctions.

The path to overcoming Th2 dominance lies in restoring the immune system's diversity and balance. Immune diversity ensures responsiveness to a wide range of challenges, while balance prevents the overactivation of any single pathway. This dynamic equilibrium relies on the interplay of Th1 and Th2 responses, regulatory T cells, and other immune components that continually recalibrate based on environmental shifts where cytokines either induce or reduce inflammation.

Th2 dominance embodies a modern challenge – a consequence

of the mismatch between our evolutionary design and contemporary living. Yet, this challenge is not insurmountable. By understanding the drivers of Th2 dominance and adopting strategies to restore immune diversity and balance, we can unlock the resilience central to optimal health.

This book lays the foundation for exploring these strategies, emphasising the importance of a dynamic, balanced immune system. With targeted interventions and a renewed focus on immune efficiency, we can empower the immune system to fulfil its extraordinary potential – adapting, defending, and thriving in the face of modern challenges.

I decided to tell the story of James and how his immune system became compromised. He is not a real person, but rather the embodiment of the sufferings of many people with debilitating health issues stemming from an imbalance in their immune system. There is hope out there for all who are experiencing similar symptoms as that of James, as this book describes his journey of learning all about his body and how he came to heal himself.

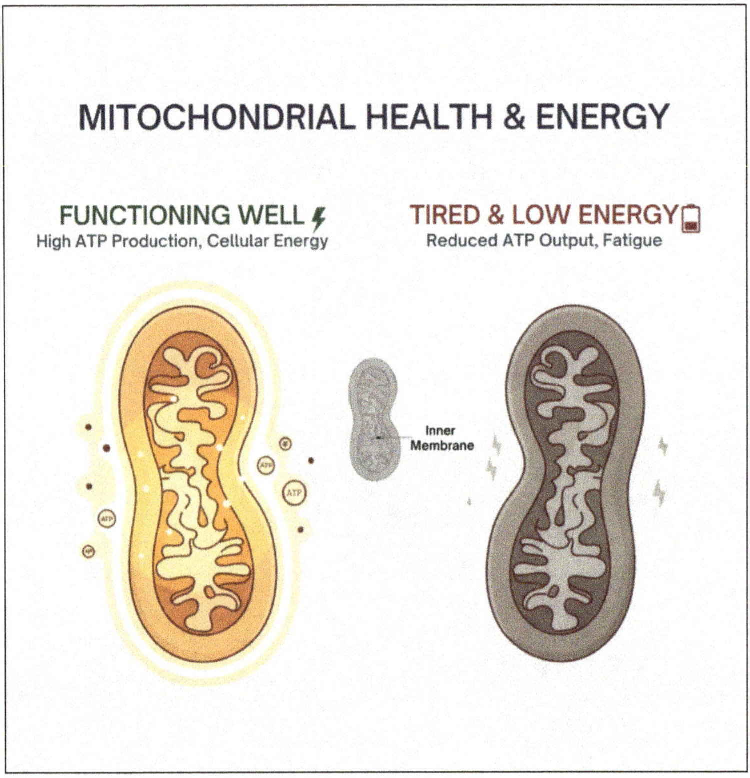
MITOCHONDRIAL HEALTH & ENERGY
FUNCTIONING WELL
High ATP Production, Cellular Energy
TIRED & LOW ENERGY
Reduced ATP Output, Fatigue
ATP
Inner Membrane

Chapter 1

Mitochondria: The Unseen Engine of Immunity and Energy

James had always assumed his constant fatigue was due to poor sleep or stress – that maybe his body just needed more rest or better nutrition. But as he began to explore the roots of his health challenges, he discovered something deeper and more surprising: the problem was not about how much energy he was consuming – it was about how much energy his body was *making*. The real culprit was not in his muscles nor his brain. It was in the microscopic engines inside nearly every cell of his body – his mitochondria.

If each cell in the body were a tiny city, the mitochondria would be its power plants. These small bean-shaped structures do not just provide energy; they manage the cell's fuel, control waste output, regulate communications between cells, and even help decide whether a cell should live or die. They are likened to a hybrid of electrical grid managers, recycling facilities, and emergency responders, all rolled into one. And like any good power plant, when they are working well, the whole city thrives. But when they begin to malfunction – even just slightly – the entire system starts to flicker and dim.

Each day, the human body produces an amount of adenosine triphosphate (ATP) – the energy currency of cells – equal to your entire body weight. Imagine having to recharge an electric vehicle with the full energy it needs to drive around the world *every single day*. That is how vital – and how hard-working – your mitochondria are. And unlike the food in your fridge, ATP cannot be stored. It must be made fresh, on demand, every second.

James' mitochondria, however, were not doing that well. Over time, they had shifted into what scientists call the *Cell Danger Response (CDR)*. This is the body's version of a lockdown protocol – like when a school goes into alert mode after a potential threat. It is a helpful, even life-saving, response in the short term. But if the system forgets to turn the alarm off, it becomes a problem.

The CDR was first described by Professor Robert Naviaux, who showed that when cells are stressed – by toxins, infections, trauma, or emotional strain – they stop cooperating and go into survival mode. Mitochondria stop making energy efficiently and begin sending danger signals instead, flooding the environment with alarm chemicals like extracellular ATP and reactive oxygen species (ROS), which are unsteady oxygen-based molecules (Naviaux, 2014).

It is as if your cell's power plant turned off the lights and started blaring sirens to warn the rest of the body.

Normally, the body would resolve this response, repair the damage, and return to normal operations. But in James' case, the danger signal never got switched off. His body had become like a town stuck in emergency mode – roads half-closed, systems running inefficiently, people exhausted from chronic alertness. And the consequences of this extended lockdown weren't just low energy – they were also changing the way his immune system worked.

This is where things became interesting – and personal.

James had always experienced allergies, skin issues, and food sensitivities. Doctors told him it was his immune system being 'overactive.' But now he began to see that these were symptoms of *misdirected* immune activity. His mitochondria, locked in survival mode, were affecting the communication lines of his immune system.

The immune system could be likened to a military with different branches: the rapid-response team (Th1 cells); the cleanup crew (T regulatory cells [Tregs]); the heavy artillery (natural killer

[NK] cells); and the border patrol (Th2 cells). In a healthy body, these branches work together with balance. But under CDR1 (the lockdown mode), the body begins to rely more heavily on the Th2 'border patrol' arm. These are the cells that help fight off parasites and regulate allergic reactions – but they are not designed to handle chronic threats such as carcinogenic cells and stealth viral infection.

With his Th1 and NK cells suppressed – essentially his heavy artillery offline – James' immune system was left with overzealous border patrol agents (Th2 cells) that began misidentifying everything from pollen to food as a foreign invader. The result? Constant allergic flare-ups, gut issues, and a fatigue that no amount of sleep could solve.

Even more concerning, his cells were starting to act like stressed workers in a power plant. When mitochondria are distressed, they start releasing mitochondrial DNA into the bloodstream – a kind of cellular 'code red' that further inflames the immune system. This feedback loop between damaged mitochondria and an overstimulated immune system is like having a broken alarm system that calls the fire department every hour, even when there's no fire. Eventually, the firefighters are exhausted, resources are stretched thin, and real emergencies go unnoticed.

This explained everything: his fatigue, his brain fog, his sluggish digestion, and his inability to bounce back from exertion. His mitochondria were stuck in a perpetual state of emergency. His immune system had followed suit, and the longer this cycle continued, the more entrenched his symptoms became.

Healing, then, was not simply about 'boosting' energy or 'supporting' immunity. It was about *retraining* his cells to feel safe again. To restore trust. To turn off the sirens and turn the lights back on.

This required a multi-layered approach: replenishing mitochondrial nutrients such as magnesium, CoQ10, and B-vitamins; reducing exposure to environmental toxins and chronic infections; and gently encouraging cellular repair through rest, breath, movement, and consistent circadian rhythms. Most of all, it meant helping his mitochondria shift out of survival mode and back into a state of energy production and cooperation.

James began to understand that his Th2-dominant immune profile – something he had once seen as a frustrating mystery – was a reflection of this deeper story. A story of cells trying to protect him, even as they lost track of the original threat. A story of energy being diverted from healing to defence. A story of disconnection that could, slowly, be rewired.

Beneath every diagnosis lies a deeper context – and at its foundation is the mitochondrion. Far more than an energy generator, it is an ancient sentinel of survival, silently influencing how the body interprets and responds to the world. When its signalling becomes distorted, the immune system follows suit, mistaking safety for threat. As we move forward, we will begin to unravel how this mitochondrial miscommunication contributes to Th2 immunological dominance – fuelling everything from allergies to immune suppression. And with this understanding, we open the door to identifying not only the hidden drivers of this imbalance – environmental exposures, nutritional insufficiencies, and hormonal cues – but also the targeted interventions through which balance may be restored. In tracing James' story, we begin to see the terrain of our own, and with it, the map towards resilience.

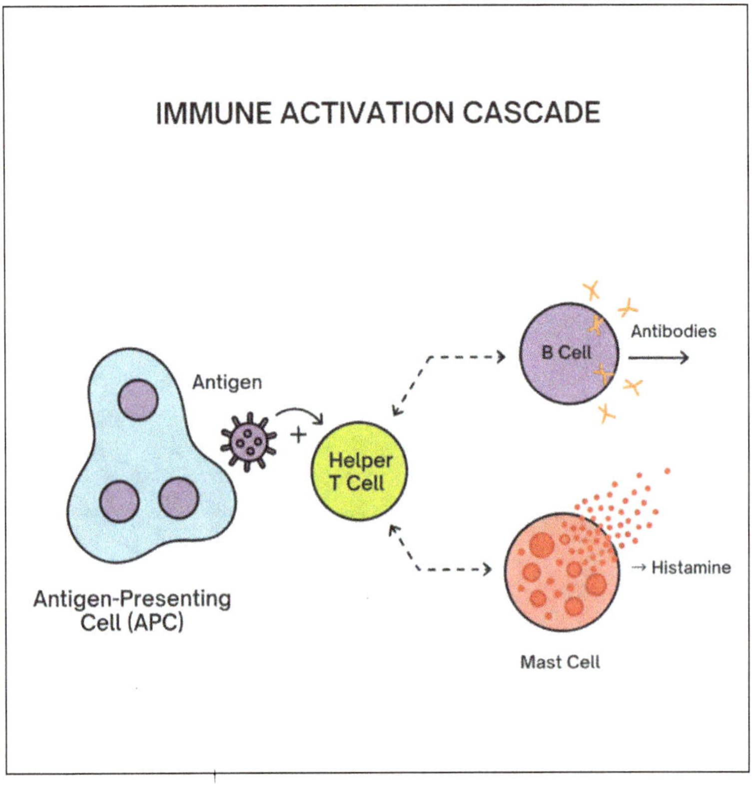
IMMUNE ACTIVATION CASCADE
Antigen
Helper T Cell
B Cell
Antibodies
→ Histamine
Mast Cell
Antigen-Presenting Cell (APC)

Chapter 2

The Language of Immunity: A Story of Th2 Responses

James had never considered himself particularly lucky or unlucky. He simply lived life, navigating its currents as best he could. But when his body betrayed him in the form of relentless allergies, recurring respiratory infections, and a hypersensitive gut, he realised that something within him was fighting a battle he barely understood. His doctors called it a Th2-dominant immune response, an imbalance in his body's defence system that had been shaping his health for years. He was about to learn that immunity is not a simple matter of strength or weakness, but of balance.

It began, as many things do, with a childhood filled with bouts of bronchitis and eczema that refused to yield to creams and inhalers. James' mother had grown used to seeing him scratch at his arms in frustration or struggling for breath on cold nights. 'He's just got sensitive skin, and his lungs are delicate,' she'd say, but deep down, she worried. No one ever spoke of how James' body seemed to react to the world as if it were a constant threat. No one explained that beneath the surface, his immune system had chosen a very specific strategy to protect him – one that came with its own costs.

These costs stemmed from an over-reliance on Major Histocompatibility Complex class II (MHC II)-mediated responses – protective, yes, but overly vigilant – priming his body to react disproportionately to otherwise harmless stimuli.

The Watchful Guardians: MHC II and the Art of Recognition

The immune system is a world of intricate communication, where cells talk in the language of molecules and recognise threats through a delicate dance of detection and response. For James, the conversation was happening at a heightened pitch. His Th2 cells, responsible for coordinating the fight against external invaders, had become overzealous protectors.

In a healthy system, immune cells rely on a set of molecules known as *MHC II* to present pieces of invading pathogens, like offering up a photograph of a suspect to a security team. This system ensures that foreign threats – such as bacteria, fungi, and parasites – are swiftly neutralised. But in James' case, this process had been set on high alert. His immune cells were constantly scanning, detecting danger even where none existed. Harmless pollen grains, dust particles, and food proteins were mistaken for enemies, triggering waves of immune activity that flooded his body with inflammation.

The First Line of Defence: Phagocytes at War

James' macrophages – the immune system's frontline scavengers – were in a perpetual state of readiness. These cells, designed to engulf and digest invading pathogens, were like sentries on a battlefield, always waiting for the next attack. Whenever he inhaled dust or was exposed to a new allergen, his macrophages acted swiftly, breaking down the intruders and sending signals to other immune cells.

Phagocytosis, the process by which the macrophage cells devour pathogens, was meant to be a protective measure, but in James' body, it had turned excessive. His immune system was not only eliminating threats; it was also setting the stage for an inflammatory cycle that left him feeling exhausted, his skin irritated, and his airways constricted. The reason for this overreaction was that a deeper immunological imbalance had taken root early in life – likely driven by a combination of genetic predisposition, early antibiotic use, and limited microbial exposure. These factors had skewed his immune system towards a Th2-dominant response, one that favours allergic reactivity and inflammation over tolerance and regulation.

An Army with Too Many Targets: IgE and the Fight Against the Invisible

One spring afternoon, as James walked through a field of blooming flowers, his eyes began to itch. His throat tightened, and a familiar burning sensation crept into his lungs. His body was reacting as though it had encountered a parasite, but there was no parasite in sight. His immune system, primed for defence, had mobilised Immunoglobulin E (IgE), the antibody most associated with allergic responses.

IgE was meant to be a specialised weapon, one that helped the body fight parasitic invaders by latching onto mast cells and triggering the release of histamine. But in James' case, IgE had become hyperactive, leading to an overreaction that played out as hay fever, asthma, and hives. The mast cells in his tissues, particularly in his skin and respiratory tract, had become like hair-trigger explosives, releasing torrents of histamine at the slightest provocation. His runny nose, his red eyes, his swollen lips – all were the result of his immune system mistaking the world around him for a battleground.

Cross-Presentation: The Double-Agent Strategy

James' immune system had another trick up its sleeve – one that made its reactions even harder to understand. Normally, part of his immune system, called MHC class II, was supposed to show pieces of outside invaders (such as bacteria or allergens) to the helper immune cells, specifically Th2 cells, which then organise a defence. But in James' case, this system sometimes went off-script: Instead of just alerting the helper cells, it also accidentally triggered the killer immune cells – called cytotoxic T cells – to join the fight. This meant that his immune system was not only responding to threats; it was amplifying the response in a way that caused more damage to his own tissues, such as his skin and lungs, making symptoms like eczema and asthma even worse.

This process, called *cross-presentation*, was meant to offer extra layers of protection, ensuring that even hidden threats – like intracellular bacteria escaping into extracellular spaces – were dealt with efficiently. But for James, it meant that some of the substances his body was reacting to weren't just being flagged for an antibody response; they were also triggering a more aggressive, dual-layered immune attack. It was as if his immune system suspected a covert enemy hiding in plain sight, treating innocent proteins as if they were masked invaders.

The Hidden Complexity: Antigen Polymorphism and the Changing Face of Immunity

James' immune system didn't just recognise threats such as antigens – it also refined its ability to identify them based on the unique shapes and sequences of invading proteins. The fragments of proteins presented by MHC II were typically between *13 and 25 amino acids* long, allowing for a vast range of immune recognition.

This flexibility ensured that the body could adapt to a wide variety of potential threats. However, it also meant that even small mutations or subtle variations in harmless proteins – such as those found in food or environmental allergens – could be enough to provoke an immune reaction. It explained why James had developed allergic responses to foods he had once tolerated as a child. His immune system wasn't just overreacting; it was adapting too aggressively, perceiving minute changes in common proteins as signs of danger.

Histamine Release: The Price of Protection

Histamine is not just a nuisance; it is an essential part of the body's defence system. It increases blood vessel permeability, allowing immune cells to reach the site of infection more efficiently. It signals the nervous system, alerting the body to potential harm. But in James' case, histamine was being released in excessive amounts, causing his body to respond with inflammation so fierce that it became debilitating.

The very mechanism designed to protect him was also harming him. His airways tightened not because of an actual invader, but because his immune system had decided that the simple act of breathing in pollen was an attack. His gut exhibited bloat, distention, and alternating bouts of diarrhoea and constipation to foods that others digested with ease, not because they were inherently dangerous, but because his body had learned to see them as threats.

The Architects of Healing: M2 Macrophages and the Scars of Inflammation

While his Th2-dominant immune system was primed for external threats, it also had another function – tissue repair. The same pathways that led to inflammation also guided healing. In the aftermath of an allergic reaction or infection, James' M2 macrophages worked to restore order. These cells were the builders, the ones that laid down new tissue, repaired wounds, and quieted inflammation.

But even healing could go wrong. When M2 macrophages were overactive, they contributed to *fibrosis* – the excessive buildup of tissue that could lead to scarring in his lungs and digestive tract. James had learned this the hard way when his gut began to show signs of inflammation and irritation that lingered long after his allergic reactions had subsided.

The Burden of an Overprotective System

James' immune system had chosen a path of defence that was as relentless as it was protective. Th2 dominance was a survival strategy, one that helped the body fight off parasites and repair wounds, but in his case, it had become an obstacle to health.

It was only when he began to see his body not as an enemy but as a misunderstood protector that he started to heal. He learned to calm the signals, to balance his immune responses through diet, stress management, and targeted therapies. His journey was not just about reducing symptoms – it was about teaching his body to trust the world again.

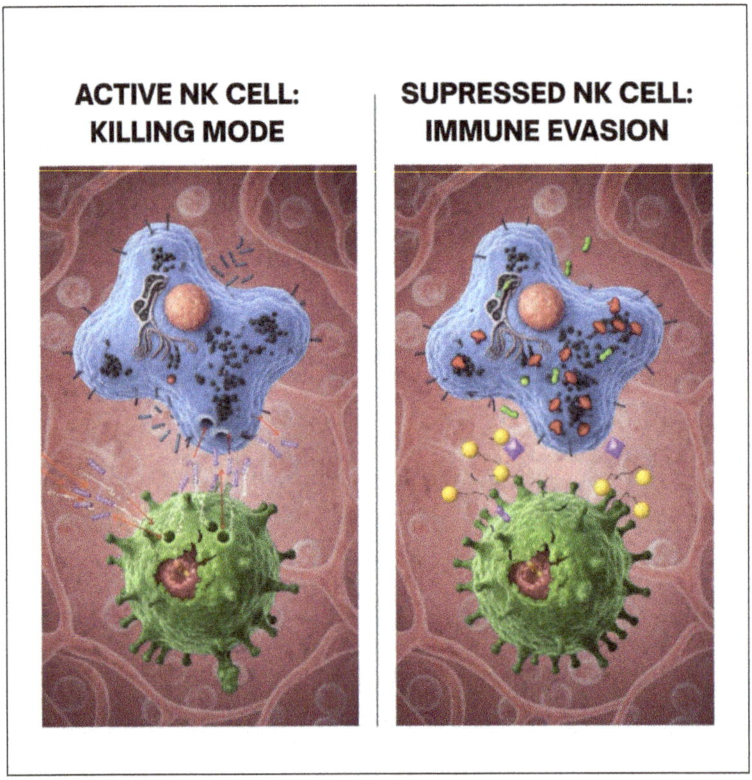
ACTIVE NK CELL:
KILLING MODE
SUPRESSED NK CELL:
IMMUNE EVASION

Chapter 3

The Silent Guardians and the Betrayal of Balance

James had always believed that the greatest battles were the ones fought in plain sight. Wars, diseases, and even personal struggles – left visible scars. But what he hadn't realised was that some of the most defining struggles of his life were unfolding in the quiet, unseen corridors of his immune system. His body had chosen a side in a battle that he did not know existed, and that choice had consequences far beyond the allergies and inflammatory responses he had grown used to. It was a betrayal not of weakness, but of overprotection, an immune system that had become too vigilant for its own good.

His Th2-dominant immune response, the force behind his relentless allergies, was shaping much more than just his reactions to pollen and food. It was silently shifting the balance of his body's natural defences, weakening the very soldiers meant to protect him from deeper, more insidious threats – his NK cells and Cytotoxic T Lymphocytes (CTLs). These cells, once his unseen guardians against infection and disease, were being systematically restrained, leaving his body vulnerable in ways he never imagined.

The Silent Warriors: Natural Killer Cells in Chains

The body is built on checks and balances, a delicate harmony of attack and restraint. NK cells are among its most formidable defenders, the silent enforcers that patrol the body, seeking out rogue cells – mutated, infected, or dangerous – before they have a chance to cause harm. They do not wait for permission. They act swiftly and without hesitation, recognising danger in a way few other cells can.

But for James, these warriors had been shackled. His Th2-skewed immune system – driven by cytokines like IL-4 and IL-10 – had suppressed NK cell activity, dampening their ability to detect and eliminate threats. It was as if the guards stationed at the gates of his fortress had been ordered to stand down, allowing intruders to slip past unnoticed.

At first, the effects were subtle. A lingering flu, infections that took long to heal, and unexplained fatigue. But beneath the surface, something far more troubling was at play. Without NK cells at full strength, his body's ability to eliminate precancerous cells before they could take root, was compromised. Every day, cells divide and replicate and, in the process, some go awry. NK cells are there to correct these errors, to seek out the ones that might turn malignant and remove them before they can cause harm. But for James, those mistakes were slipping through, unchallenged.

The realisation was unsettling. It meant that his immune system, designed to protect him, was inadvertently increasing his risk of cancer. The overactivity of Th2 had created a paradox – an immune response that was too strong in the wrong ways and too weak in the places that mattered most.

The Weakened Assassins: Cytotoxic T Cells Under Siege

While NK cells act as the first line of defence, CTLs are the immune system's specialised assassins. They are precise, strategic, and relentless, seeking out and eliminating cells infected with intracellular pathogens – viruses and bacteria that hide within the body's own cells, and even early-stage tumour cells.

For James, however, the orders had changed. His Th2-dominant immune system had not just weakened his NK cells but had also stripped his CTLs of their weapons. Normally fuelled by cytokines like interferon-gamma (IFN-γ) and interleukin-2 (IL-2), which empower them to attack, his CTLs were starved of the signals they needed to function. Without these essential cues, they remained in a dormant state, unable to mount an effective response.

Viruses that would have once been swiftly neutralised now lingered longer, their presence tolerated rather than fought. Bacteria that should have been eradicated found opportunities to spread. And perhaps most dangerously, the very cells in his body that could turn cancerous were no longer under surveillance.

It was an eerie parallel to the way his body mistook harmless substances like pollen for threats, while ignoring the dangers that lurked unseen. His immune system had become an unreliable narrator, spinning tales of danger where none existed, while turning a blind eye to the real threats growing in the shadows.

The Cancer Conundrum: When the Body's Defenders Falter

The impact of Th2 dominance on NK cells and CTLs was not just an abstract concern – it had real-world consequences. Cancer immunologists had long recognised that individuals with strong Th2-skewed immune responses were at a higher risk of developing malignancies. The reason was clear: with weakened NK and CTL activity, the body's ability to detect and eliminate precancerous cells was compromised. Tumours could grow unchecked, their presence unnoticed until it was too late.

For James, this realisation hit hard. He had spent years battling an immune system that overreacted to the harmless, never considering that it might also be underreacting to the dangerous. The very imbalance that left him breathless in the spring air was also one that could allow something far more sinister to take hold.

But this wasn't just about risk – it was about strategy. Scientists and physicians have been exploring ways to restore balance, to reinvigorate NK cells and CTLs in individuals like James. Some of these approaches involve immune checkpoint inhibitors, designed to remove the brakes that keep CTLs in a weakened state. Others focus on modulating Th2 responses, shifting the immune balance back towards a more functional equilibrium, which will be detailed in later chapters.

The answer, it seemed, was not about choosing between strength and weakness, but about restoring harmony.

Restoring the Balance: A Path Forward

James knew that changing his immune system would not be easy. Years of Th2 dominance had left deep imprints on his body, patterns that would not simply vanish overnight. But he also knew that the body was adaptable, capable of learning and adjusting when given the right tools.

His journey towards restoring balance would extend beyond medication alone, incorporating diet, targeted nutritional interventions, and stress management techniques. Nutrient-rich meals featuring anti-inflammatory foods – such as blueberries (rich in anthocyanins that neutralise oxidative stress), salmon (containing omega-3 fatty acids EPA and DHA, which reduce inflammation), kale (high in vitamin K and carotenoids, supporting immune cell function), and turmeric (providing curcumin, known to inhibit inflammatory signalling) – could meaningfully regulate and strengthen his immune response. Complementing these dietary adjustments with stress-reduction practices, like meditation and breathwork, would further enhance immune balance and overall well-being.

Most importantly, he had to change his perspective. His immune system was not his enemy – it was simply misdirected. By understanding its language, and by listening to its signals without fear, he could guide it back to a place of equilibrium.

James had spent years fighting his body. Now, he was ready to work with it – beginning a journey to heal his immune system, which is explored in detail later in this book.

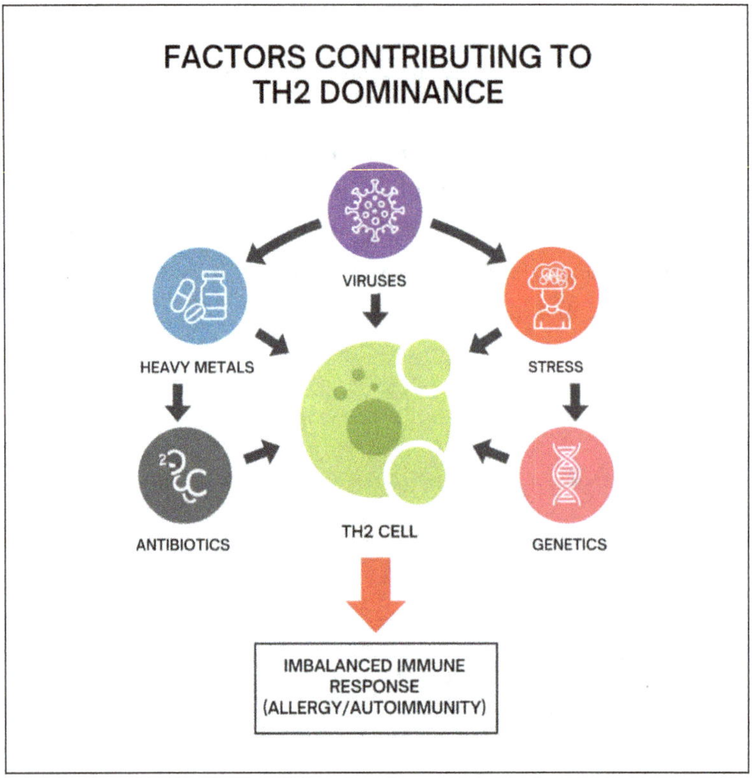
FACTORS CONTRIBUTING TO
TH2 DOMINANCE
VIRUSES
HEAVY METALS
STRESS
ANTIBIOTICS
TH2 CELL
GENETICS
IMBALANCED IMMUNE
RESPONSE
(ALLERGY/AUTOIMMUNITY)

Chapter 4

The Forces That Shape Immunity: Understanding Th2 Dominance

James had always suspected that his immune system wasn't just reacting – it was learning. Every wheeze, every rash, every lingering bout of inflammation seemed to tell a story of adaptation, of an internal landscape shaped by forces beyond his immediate control. He had once thought of his immune system as an unwavering shield, an armour forged in childhood. Now, however, he recognised it as an intricate web of responses – an evolving interplay of genetics, lifestyle, and environmental influences that had led him towards Th2 dominance.

As he dug deeper into his condition, he realised that his Th2-dominant immune response was not a random malfunction. It had been sculpted by a lifetime of exposures – things he had breathed in, swallowed, absorbed, and endured. Some factors had directly pushed his immune system towards this hyperactive state. Others had simply increased the odds, nudging him further along a path of reactivity. Together, they told a story of how immunity is shaped not just by biology, but by history – his own, and that of the world he lived in.

A Delicate Balance Disrupted: The Hidden Architects of Th2 Dominance

The Invisible Hand of Antibiotics

James could still remember the countless bottles of pink syrupy medicine his mother had given him as a child. Every winter brought another round of infections, and with them, another dose of antibiotics. They had seemed like a miracle then – until he began to wonder what they had left behind.

Antibiotics were meant to kill bacteria, but they did not discriminate between friend and foe. With every prescription, his gut microbiome – the intricate ecosystem of bacteria that helped regulate his immune system – was altered: the antibiotics killed the bad bacteria as well as the good bacteria. The balance shifted. The bacteria that once trained his immune cells to tolerate harmless invaders began to dwindle. And without this early conditioning, his immune system lost some of its ability to distinguish between a true pathogen and an innocent allergen. It overreacted. It saw shadows where there were none.

Scientists had found that this wasn't just a coincidence. Antibiotic overuse in childhood had been linked to high rates of allergies, asthma, and autoimmune diseases. The tiny organisms that once played a crucial role in immune education were disappearing, leaving behind a system primed for unnecessary battle.

The Chemical Burden: Persistent Organic Pollutants

James had never considered how much of the modern world his immune system had been forced to absorb. Chemicals leaching from plastic packaging and water bottles, pesticides such as organophosphates sprayed on genetically modified crops, and industrial chemicals lingering in household cleaners and the air he breathed – these had all found their way into his body, quietly shaping his biology towards Th2 immunological dominance.

Persistent organic pollutants (POPs) were among the most insidious chemicals that James encountered. Resistant to degradation, they silently accumulated in his tissues, whispering distorted instructions to his immune cells through molecular mimicry. These compounds – particularly dioxins (highly toxic by-products found in herbicide production, paper bleaching, and combustion emissions) and polychlorinated biphenyls (PCBs) (industrial chemicals historically used in paints, glues, electrical transformers, capacitors, and even lingering in farmed salmon and contaminated waterways) – interfered with cytokine signalling. Normally, this chemical messaging kept his immune system balanced, but in James' body, the persistent presence of these pollutants tipped the scales towards Th2 dominance, suppressing the pathways that usually kept allergic responses under control.

He read about studies showing that dioxins altered immune function so profoundly that they could shift an entire population towards increased allergy and autoimmune risk. It made sense. He wasn't just reacting to his environment – his environment had already written the script.

The Toxicity of Heavy Metals

There was another unseen force at play, one that lurked in ageing water pipes, in industrial runoff that contaminated rivers and drinking water, and even in the polluted air of cities. Heavy metals – lead, mercury, and cadmium – were silent saboteurs of immune function. These toxic elements seeped silently into cells, interfering with signalling pathways and warping immune responses.

James had stumbled upon the story of thimerosal, a mercury-based preservative once used in vaccines and medical products. Research suggested that it could shift the immune system towards a Th2-dominant state by altering dendritic cells – the specialised immune cells responsible for capturing foreign substances, breaking them down, and presenting them (via molecules called MHC) to T cells, effectively teaching the immune system what to attack and what to tolerate. Instead of training his body to efficiently combat viruses and bacteria, these dendritic cells had become misdirected, nudging his immune response towards a hyperactive, allergy-prone state.

Lead, too, had a story to tell. Exposure to this heavy metal was known to weaken the immune defences that relied on Th1 activity, the branch of the immune system responsible for targeting viruses and bacteria hidden inside cells. Th1 responses typically act as a vital counterbalance to Th2-driven allergic and inflammatory reactions. The more lead someone was exposed to, the less capable their immune system became at effectively eliminating these intracellular infections, increasing the likelihood that it would instead overcompensate with inflammatory, Th2-driven responses. James couldn't help but wonder: how much of his immune system's misfiring had been dictated by things he had unknowingly absorbed?

The Shadows of the Past: Viral Triggers and the Hygiene Hypothesis

Viruses That Refuse to Leave

James had never given much thought to the illnesses that had come and gone over the years. A fever, a sore throat, a week in bed – he always recovered. But what if those infections had never truly left?

Some viruses, like Epstein-Barr virus (EBV), didn't simply pass through. They lingered, hiding inside cells, evading immune detection. EBV was notorious for manipulating the immune system, suppressing the very responses that should have kept Th2 activity in check. It released its own version of interleukin-10 (IL-10), a molecule that helped it evade attack but also reinforced Th2 dominance in the process.

James' history of fatigue, inflammation, and recurring allergies now seemed to make more sense. Not only was his immune system reactive – it was also being manipulated by invisible forces.

Hygiene – Too Clean for Our Own Good?

Then there was the hygiene hypothesis. It suggested something almost paradoxical: that modern cleanliness was contributing to the rise of allergies and autoimmune conditions. James' childhood, spent in a meticulously sanitised environment, had left him without the microbial exposures that earlier generations had encountered daily.

His immune system, starved of real threats, had turned against harmless invaders. Without the dirt, the bacteria, and the early infections that might have trained his body to tolerate the world, it had learned to overreact instead.

Studies had shown that children who were raised on farms and regularly exposed to animals and soil, had significantly lower rates of allergies and autoimmune diseases (Riedler et al., 2001). Their immune systems benefited from a diverse microbial landscape, helping them develop a more balanced response to environmental triggers. James, like so many others, had grown up in an era of antibacterial soaps, processed foods, filtered water, and air-purified homes – and in that overly sterile world, his immune system had been left searching for something to fight.

The Internal Factors: Oestrogen and Atopic Conditions

James had also begun to see how biology itself played a role in his immune story. His mother and his sister had both struggled with allergies and autoimmune issues. Women, he learned, were more likely to develop Th2-dominant conditions, largely due to the influence of oestrogen.

Oestrogen was not just a reproductive hormone; it was an immune modulator. High levels promoted Th2 cytokine activity, making allergic and autoimmune responses more likely. It was why women experienced high rates of conditions like lupus and rheumatoid arthritis. The very hormones that supported life also played a role in the immune system's miscalculations.

Atopic conditions – such as eczema, asthma, and hay fever – were yet another clue. They were all linked to Th2 dominance and were part of the same biological pattern. His history of eczema as a child had been a warning sign, a glimpse into an immune system predisposed to misinterpretation.

Rewriting the Immune Narrative

James had always thought of his allergies, his sensitivities, and his immune responses as personal failures – things to be battled and suppressed. But now he saw them for what they truly were: adaptations that were shaped by his world, his experiences, and his biology. His body wasn't broken; it had simply learned the wrong lessons. And lessons could be unlearned.

There was no single solution, no quick fix. But understanding the forces that had led him here meant he could begin to undo their effects. He could rebuild his microbiome, reduce his exposure to pollutants, support his detoxification pathways, and find ways to retrain his immune system to see the world with less suspicion.

For the first time, James felt something new – something powerful.

Hope.

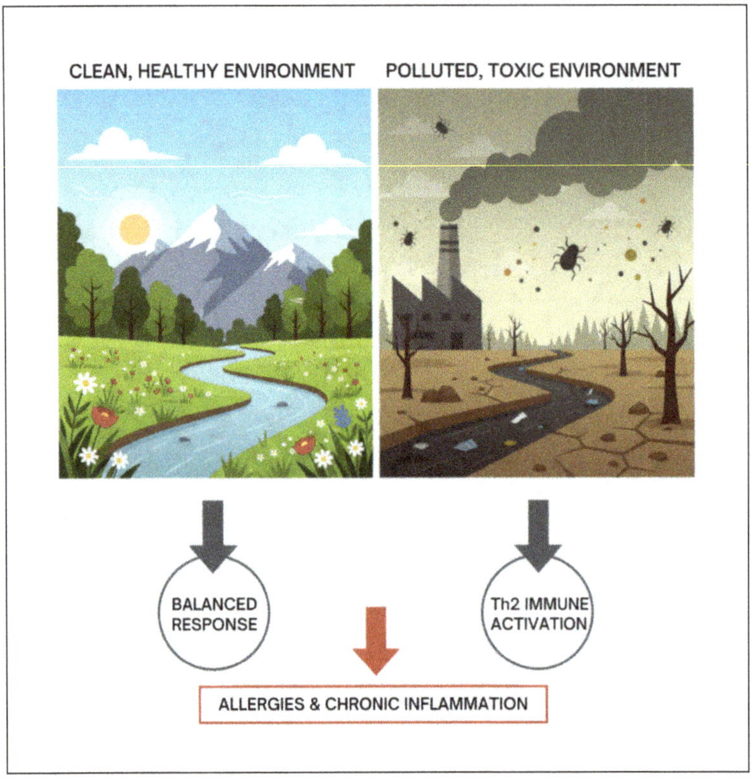
CLEAN, HEALTHY ENVIRONMENT
POLLUTED, TOXIC ENVIRONMENT
BALANCED RESPONSE
Th2 IMMUNE ACTIVATION
ALLERGIES & CHRONIC INFLAMMATION

Chapter 5

The World Around Us: Environmental and Additional Factors in Th2 Dominance

James had always thought of his immune system as an internal problem – something malfunctioning deep within his biology. But the more he learned, the more he realised his immune system wasn't acting alone. It was responding to the world around him, shaped by his environment as much as by his genetics.

His hypersensitivities were not just the result of a misfiring defence system – they were a conversation between his body and his surroundings, one that had started long before he was aware of it. From the air he breathed to the stress his mother had carried before he was born, every element played a role in shaping the Th2-dominant response that now governed his immune system.

Understanding these influences was not just about science – it was about rewriting the relationship between himself and the world.

A World That Demands a Response

The Ubiquity of Allergens: How the Body Learns to Overreact

James' first memories of spring were filled with beauty – the vibrant green of new leaves, the scent of flowers carried by the wind. But alongside that beauty came the discomfort: itchy eyes, a burning throat, and the unmistakable feeling of congestion settling in. His body saw the season's arrival as an invasion.

Pollen, dust mites, and pet dander were not supposed to be threats. For most people, they were harmless environmental particles, encountered and ignored by the immune system. But for James, these substances triggered a cascade of immune activity. His mast cells, hypersensitive and overpopulated with IgE antibodies, stood at attention like nervous sentries, primed to unleash a flood of histamine at the first sign of exposure.

It was not just about genetics – it was about repeated exposure shaping his immune system's memory. Every allergic reaction reinforced the body's decision to treat these benign substances as enemies, strengthening the very responses that made his symptoms worse.

Climate: The Environmental Architect of Immune Dysregulation

Cold, dry winters altered the very structure of the body's defences. The skin, meant to be the body's first line of protection, lost its moisture and cracked under the pressure of low humidity. James' struggles with eczema and seasonal asthma now made sense to him.

Humid climates can also be problematic. In tropical and urban areas, where pollution combines with high humidity, allergens bind themselves to fine particulates in the air, making them even more potent. For some people, these environmental shifts can trigger the first signs of atopy, subtly pushing their immune systems towards chronic hypersensitivity.

The skin was not just a passive barrier; it was a crucial part of immune function. When it was weakened, allergens and irritants had a direct pathway into his system, triggering the Th2-dominant response that had already been primed by his genetics and environment.

Pollution: The Silent Trigger of Th2 Dominance

James recognised that air pollution not only affected his lungs, causing coughing and respiratory irritation – it was also shaping his immune system.

The fine particulates from car exhaust, industrial smoke, and household chemicals are physical irritants that disrupt immune signalling, increasing oxidative stress. Oxidative stress occurs when the free radicals caused by the pollutants trigger an inflammatory response, – tipping the balance towards Th2 dominance. Studies have shown that cities with higher levels of pollution also have higher rates of asthma and allergies – not just because people are inhaling pollutants, but because these pollutants are actively reshaping their immune function.

Then there was tobacco smoke – an immune paradox. In low exposures, it could trigger allergic responses, such as inflamed airways, asthma symptoms, or heightened sensitivity, driving the body towards IgE-mediated inflammation. But in high doses, nicotine has been found to suppress Th2 immune response. This immunosuppressive effect was often observed in populations with high nicotine consumption, particularly in regions where parasitic infections were also more common. Some researchers had speculated that the body, in adapting to chronic nicotine exposure, became more resistant to parasitic infections while simultaneously dampening allergic responses.

James learned that his immune system was constantly recalibrating in response to external forces he was unaware of.

Internal Influences: When the Body Works Against Itself

A System Primed for Overreaction

James' immune system didn't just respond to the outside world – it also struggled with its own internal balance.

At the heart of his reactions was an overabundance of IgE, the antibody responsible for allergic responses. IgE was not supposed to be this active. In a healthy immune system, it was reserved for parasitic infections, helping the body mount a defence against worms and other invaders. But in James, IgE had become a trigger-happy executioner, launching inflammatory assaults on harmless substances.

The problem was not the IgE itself – it was the failure of his regulatory immune pathways to keep it in check. Without the proper balance of Th1 cells that target intracellular pathogens and Treg cells that regulate or suppress the immune response, his body had no mechanism for dialling down its responses.

A Broken Barrier: The Role of Skin Integrity

His immune dysfunction wasn't just happening inside – his skin also played a crucial role.

People with atopic dermatitis often had mutations in a protein called filaggrin, essential for maintaining the skin's barrier function. Without it, the skin became more permeable, allowing allergens to penetrate deeper and trigger immune reactions.

James had experienced this firsthand. Every winter, his skin cracked, and his eczema flared, creating a direct pathway for allergens to enter and activate his already hypersensitive immune system.

Maternal Influence and Early-Life Factors

Stress Passed Down: The Impact of Maternal Cortisol

Even before he was born, his immune system had been shaped by forces beyond his control. Maternal stress during pregnancy has been linked to Th2-skewed immune responses in offspring. Increased cortisol levels in the mother can influence foetal immune programming, priming it for hypersensitivity and allergic conditions.

Had his mother been stressed while carrying him? Had her financial and relationship worries, her anxieties, or her physiological responses left an imprint on his developing immune system?

Science suggested that the answer was likely yes.

Food as an Early Teacher

Then there was his diet – one of the earliest ways the immune system learned to distinguish between friend and foe.

Breastfeeding had been shown to reduce the risk of atopic conditions, providing critical immune-modulating compounds that help train the body's defences. Early exposure to diverse foods also played a role; introducing common allergens like peanuts and eggs at the right time could reduce the risk of developing allergies later in life.

For James, this raised new questions: Had his introduction to food contributed to the shaping of his immune responses? Could dietary interventions help recalibrate them now?

Reclaiming Balance: The Path Forward

James was beginning to understand that his immune system was simply responding to a world that had shaped it in ways he was previously unaware of.

If his immune dysfunction had been trained by his environment, then perhaps, through careful changes, it could be retrained. By reducing pollutant exposure, rebuilding his microbiome, managing stress, and supporting skin integrity, he could begin to shift his immune responses towards a healthier balance.

The world had shaped him. Now, it was time for him to shape his world.

DENTAL INFECTION & SYSTEMIC INFLAMMATION
Inflammatory Signals Travel Through The Body

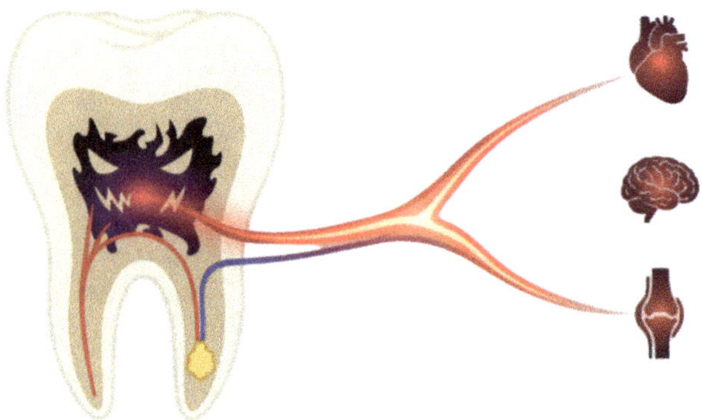

Inflammatory Mediators

Chapter 6

The Forgotten Gateway: Oral Health, Silent Inflammation, and the Roots of Immune Dysregulation

James had long thought that true healing was a matter of diet, movement, and stress management – the big three that most wellness paths preached. But as he dug deeper into the hidden architecture of his immune system, he discovered a neglected domain, one that doctors hardly ever mention: his mouth.

It seemed absurd at first. Dental checkups were about cavities, braces, or the odd root canal. How could oral health be relevant to the fatigue, the inflammatory fog, or the immune imbalance that governed his life?

And yet the more he read, the more undeniable it became: it all started in the mouth.

James stumbled upon the work of Dr Dominik Nischwitz (also known as Dr Dome), a pioneer in biological dentistry, whose book *It's All in Your Mouth* framed the oral cavity not as an isolated system but as the gateway to metabolic, neurological, and immune resilience. Chronic illness, Nischwitz insisted, could not truly be reversed without restoring oral health first – a fact that was never considered in any of James' previous treatments. (Nischwitz, 2020).

The mouth is a highly vascularised immune interface, directly linked to the gut, the sinuses, the brain, and the systemic circulation.

Every root canal, improperly healed extraction, or silent infection, carries the potential to sabotage immune regulation.

James' curiosity led him deeper into the research – particularly studies by Dr Johann Lechner and colleagues – that revealed an astonishing phenomenon: silent jawbone cavitations, clinically termed Fatty Degenerative Osteonecrosis of the Jaw (FDOJ) (Lechner et al., 2024). These cavitations often developed after tooth extractions, root canals, or poorly healing dental surgeries. Instead of regenerating healthy bone, the jawbone developed areas of fatty degeneration, ischemia (inadequate blood supply), and chronic 'silent' inflammation (Lechner and Von Baehr, 2017).

Unlike typical infections, these cavitations produced few, or no, symptoms of throbbing pain or overt swelling. Standard X-rays often missed them altogether, offering false reassurance. Yet within these seemingly quiet pockets, the immune system was being persistently misdirected.

The culprit behind this misdirection was a cytokine called RANTES/CCL5 – a chemical messenger vastly overexpressed in FDOJ lesions. Where normal jawbone tissue expressed minimal RANTES, cavitation sites showed levels of up to thirty-fold higher (Lechner et al., 2024). This chronic signal subtly but relentlessly provoked the immune system towards a Th2-dominant inflammatory state.

James began to trace his history backward, peeling away the layers of misunderstanding that had framed so much of his life. As a boy, he had undergone the usual dental rites of passage – extractions of his crowded teeth, a series of fillings, and later, the removal of his wisdom teeth. Each procedure had seemed unremarkable at the time, mere inconveniences on the path to adulthood. Yet somewhere along the way, a different kind of wound had been left behind – FDOJ.

A particular memory surfaced – one he had long buried over the years. He was sixteen when it happened: a sharp crack in his lower molar during a rugby match, followed by days of aching pain and an urgent dental appointment that ended with a root canal.

At the time, it had seemed a small victory. The pain had vanished, and the tooth was saved. But looking back now, armed with the knowledge he had gained, James understood the price that had been quietly exacted.

The root canal had preserved the outer form of his tooth, but the living tissues inside – the vital blood vessels and immune defences – had been stripped away. In their absence, a hollow structure remained, providing a scaffold vulnerable to colonisation by anaerobic bacteria. These microbes, protected from the body's immune surveillance, and typically undetectable by standard dental X-rays, thrived in spaces where conventional imaging would usually reveal only structural abnormalities rather than microbial presence. From this hidden position, they released volatile compounds and immune-activating toxins into his system, silently fanning the flames of innate immune activation.

He read how researchers like Dr Nischwitz described the cytokine cascade that such hidden infections could sustain. Early bursts of innate immune messengers like tumour necrosis factor-alpha (TNF-α) and interleukin-1 beta (IL-1β) would have initially tried to defend him. But over time, this unresolved inflammation shifted the landscape of his immunity – leading to chronic overexpression of chemokines like RANTES/CCL5 and fuelling a systemic tilt towards Th2 dominance (Nischwitz, 2020).

The tooth had not merely died; it had become a silent conductor of chaos, leaking signals of distress into his bloodstream, his gut, and his brain.

His depression, his fatigue, his relentless inflammation – these were not random misfortunes. They were the echoes of old wounds, unhealed because they were unseen.

Reading further into Dr Nischwitz's work, James found the words that crystallised his suspicion. Nischwitz described how conditions like Neuralgia-Inducing Cavitational Osteonecrosis (NICO) and

FDOJ could persist for years beneath the surface, producing no obvious dental symptoms, yet unleashing a flood of inflammatory cytokines into the body's systems. These silent killers, he explained, were often overlooked in standard medical evaluations, yet they fed chronic immune dysregulation, neuroinflammation, and emotional instability (Nischwitz, 2020).

The depression he had battled for most of his adult life – the waves of fatigue, the inexplicable anxiety, the hollow fog that no amount of motivation could dispel – had not been merely psychological – it was biological. His jaw, injured and unseen, had been quietly poisoning his resilience for decades.

It was humbling, devastating, and liberating all at once. The story of his suffering was not one of personal failure – it was a story of hidden wounds, of missed connections. And now, for the first time, he could see where healing must begin.

Not only in his mind. Not only in his gut.

But in the bones of his face, where silent inflammation had built a prison from within.

He began to piece the puzzle together. Silent inflammation in the jawbone didn't cause dramatic crises; it acted like background noise – keeping the immune system in a chronic state of misfired vigilance. Persistent Th2 activation drove allergies, histamine responses, and autoimmune tendencies, while suppressed Th1 function weakened his body's antiviral and antitumour defences. Sustained RANTES signalling, therefore, promoted neuroinflammation, mitochondrial dysfunction, and metabolic slowdown (Lechner et al., 2024).

Even more alarming, studies have linked FDOJ and NICO to a range of systemic illnesses such as breast cancer, chronic fatigue syndrome, autoimmune disorders, atypical facial pain, and neuro-degenerative diseases (Lechner and Von Baehr, 2017).

Healing, James realised, was not about endlessly treating symptoms downstream. It was about finding – and removing – upstream sources of immune confusion. Silent cavitations and infected root canals were not the cause of every health problem, but for someone whose immune system was already tilted towards Th2 dominance, chronic oral inflammation acted like gasoline on a slow-burning fire.

Resolving these hidden infections wasn't a fringe idea; it was foundational. As Dr Nischwitz put it: 'The mouth is the entrance to the gut, to the immune system, and to the brain. If you want to heal the body, you must first heal the mouth.'

James felt the weight of that truth – heavy, humbling, but freeing. Another door had opened on his path to recovery. And as he stepped through it, he understood something that now seemed almost obvious.

It really did all start in the mouth.

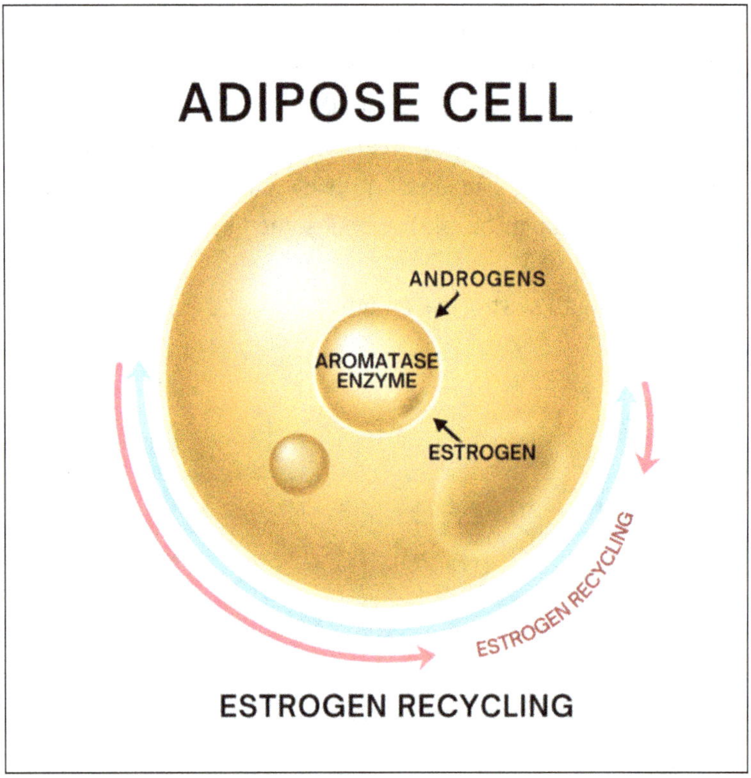
ADIPOSE CELL
ANDROGENS
AROMATASE ENZYME
ESTROGEN
ESTROGEN RECYCLING
ESTROGEN RECYCLING

Chapter 7

The Weight of Hormones: How Body Fat Shapes Oestrogen's Influence

James had always known he was carrying a few extra pounds, but like many, he hadn't thought of fat as anything more than a storage issue – something to be burned, counted, or worked off at the gym. It wasn't until his health began to unravel in more subtle ways – mood swings, fatigue, skin issues, and an even deeper shift in his immune resilience – that he started asking different questions.

Was his body fat doing more than just sitting there?

The answer, he discovered, was far more complicated than he had imagined. Fat, as it turned out, was not passive at all. It was alive with hormonal activity – a biochemical factory quietly shaping his immunity, his mood, and even how his genes expressed themselves. Beneath the surface of his skin, a silent transformation had been taking place – one that tipped his internal balance towards a hormone that had grown far too loud: *oestrogen*.

Adipose Tissue: The Endocrine Organ No One Talks About

At first, James thought oestrogen dominance was something only women needed to worry about – particularly during menopause. But then he learned that *adipose tissue* (his own body fat) was producing oestrogen around the clock, converting androgens into oestrogens via an enzyme called *aromatase*. And the more fat he carried – especially around the abdomen – the more active aromatase became.

In postmenopausal women, adipose tissue becomes the main producer of oestrogen. But even in men and younger women, elevated fat stores meant that a continuous stream of oestrogen entered the bloodstream, quietly shifting the delicate hormonal ratios that influenced everything from immune tone to neurotransmitter activity.

This was not only about having too much oestrogen – it was about losing balance. When oestrogen levels became too high compared to progesterone or testosterone, it caused inflammation, overwhelmed his body's natural detox processes, and negatively affected his mood, as James had frequently experienced.

Genes and Enzymes: The Hidden Scripts of Oestrogen Metabolism

As James dove deeper, he found another piece of the puzzle – his *genetic blueprint.* It wasn't just about how much oestrogen his fat cells were producing – it was also about how his body metabolised and cleared that oestrogen.

CYP1A1 and CYP1B1: Pathways of Possibility and Risk

His body used enzymes – such as those encoded by *CYP1A1* and *CYP1B1* – to convert oestrogen into different metabolites. Ideally, *CYP1A1* would guide oestradiol down a gentler path, producing *2hydroxyestradiol*, a relatively mild by-product. But *CYP1B1*, particularly when overactive or genetically altered, could shift oestrogen metabolism towards *4hydroxyestradiol* – a more reactive form that is capable of generating oxidative stress and damaging DNA.

James carried a variant in CYP1B1 that made this risky route more likely. His oestrogen wasn't just elevated – it was being reshaped into a molecule with inflammatory and carcinogenic threat.

The COMT Bottleneck

Next came COMT (catechol-O-methyltransferase) – an enzyme that helps break down and remove these hormone by-products through a process called methylation, where the body attaches a small molecule to toxins and hormones, making them easier to eliminate. If methylation slowed down, often due to a common genetic variation (COMT Val158Met), it created a bottleneck. This meant hormone by-products, especially those from oestrogen, could build up, causing cell damage, mood swings, and symptoms linked to excess oestrogen such as fatigue, anxiety, difficulty losing weight, reduced libido, and increased fat around the midsection – symptoms that James had frequently experienced.

James' history of anxiety and brain fog suddenly took on new meaning. These were not random experiences – they were symptoms of a detox system struggling to keep pace with hormonal demand.

Sulfotransferase (SULT) Enzymes and the Sulphation Backlog

Lastly, his body attempted to clear oestrogen through *sulphation*, tagging it for elimination with the help of *SULT1A1* and *SULT1E1*. But these enzymes, too, could be impaired – whether due to genetics, toxin overload, or nutrient depletion. When sulphation faltered, oestrogen remained active longer than it should, continuing to stimulate tissue, alter immune signalling, and overload detox systems.

It became clear: every pathway designed to manage oestrogen had the potential to falter, especially in a system already overwhelmed by fat accumulation and stress.

The Stress Connection: Cortisol, Neurotransmitters, and the Oestrogen Loop

James had always carried stress like a second skin. Work pressure, emotional suppression, and the quiet, gnawing weight of responsibility, had worn grooves into his physiology. For a long time, he thought stress just made him tired or distracted. But now he was learning how deeply it affected his hormones. He had heard of the term *'cortisol steal'*, the idea that stress redirected hormonal resources away from progesterone to feed cortisol production. But what was really happening was more subtle: *chronic stress dysregulated his endocrine priorities*.

His adrenal glands weren't robbing one hormone to make another. Instead, prolonged cortisol elevation *altered the hormonal conversation*, suppressing ovulation in women or reducing adrenal progesterone output in men. This meant that progesterone – the natural counterweight to oestrogen – was no longer available in sufficient amounts. And with low progesterone, oestrogen's excitatory, growth-promoting effects became unbalanced.

Furthermore, oestrogen was now influencing his brain. It heightened his sensitivity to dopamine, serotonin, and norepinephrine – but without progesterone's calming influence, those neurotransmitters behaved erratically. *Mood swings*, *restlessness*, and *emotional overwhelm* were not signs of personal weakness. They were echoes of a hormonal system out of sync with itself.

James realised that his hormones weren't just reacting to life's demands – they were carrying the weight of *his unprocessed experiences*.

The Gut-Thyroid Axis: Hidden Amplifiers of Hormonal Chaos

A Sluggish Thyroid and the sex hormone-binding globulin (SHBG) Drop

Another thread emerged – his *thyroid function*. For years, James had been told his labs (laboratory tests) were 'borderline', but this was never treated. Now he discovered that even mild *hypothyroidism* (underactive thyroid gland) could lower levels of *SHBG*, the protein that binds oestrogen and limits its activity.

With lower SHBG, more *free oestrogen* circulated through his body – triggering tissue responses, mood changes, and immune shifts far beyond what the total hormone levels might suggest.

The Gut's Recycling Programme Gone Wrong

And then there was the gut.

His liver worked hard to process oestrogen, packaging it for excretion. But in the intestines, a specific group of bacteria produced an enzyme called *beta-glucuronidase*, which could *uncouple* these oestrogens and send them right back into circulation – a process known as *enterohepatic recycling*.

James' gut, already compromised by a lifetime of processed food and antibiotic use, had become a *recycling plant for oestrogen*. Rather than leaving the body, hormones were being reabsorbed – retriggering the very symptoms he had tried to escape.

The solution was not only to 'balance hormones'; it was to restore the entire *ecology of elimination* – from the liver to the gut to the microbial residents that had been unknowingly contributing to his hormonal overload.

A Body Asking to Be Heard

James' weight wasn't just about calories – it was about communication. His fat stores, his stress, his genetics, and his gut – were all speaking, shaping an internal landscape where *oestrogen dominance* wasn't just a lab value, but a lived experience.

And while he couldn't rewrite his genetic code or erase his past stress, he could begin to respond differently. Through *nutrition, movement, gut healing, liver support*, and *stress integration*, he could start teaching his body a new way to respond.

Oestrogen dominance was not a diagnosis. It was a *signal* – a whisper from the body that something had been out of balance for a long time. Now, James was finally ready to listen.

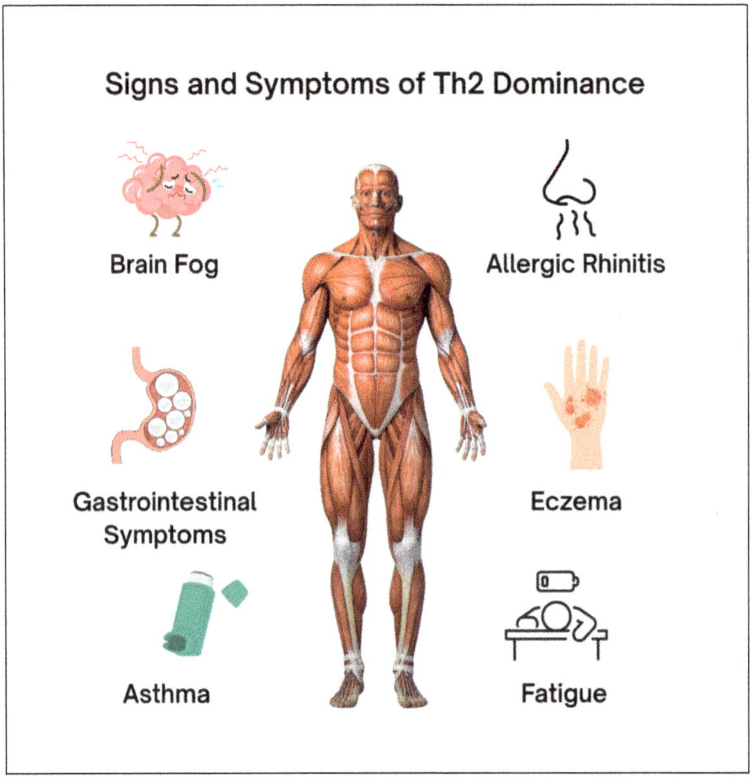

Signs and Symptoms of Th2 Dominance
Brain Fog
Allergic Rhinitis
Gastrointestinal Symptoms
Eczema
Asthma
Fatigue

Chapter 8

When the Body Speaks: Signs and Symptoms of Th2 Dominance

James had always sensed that something was not quite right. Not in the dramatic, diagnostic sense, but in the quiet accumulation of seemingly minor disruptions – an itch that never fully resolved, a fog that never cleared, a fatigue that no sleep could touch. For years, he had dismissed these as normal signs of ageing, being stressed, and of living in a fast-paced world. But now, as he explored the intricate dance of his immune system, he began to see these symptoms as signals – each one a message shaped by a dominant Th2 immune response.

It began with his skin. What he once called 'sensitive' had become reactive. A scratch would flare into a welt, and certain soaps or fabrics would cause his arms to burn, forming red patches. The return of eczema, which had faded in his twenties, surprised him – not just the dryness, but the underlying heat, the sense that his skin was under siege. As he would later learn, this was not simply a dermatological issue, but a hallmark of Th2 dominance. The elevated interleukins IL-4, IL-5, and IL-13 were impairing his skin barrier and triggering a cascade of histamine release from overactive mast cells. His immune system, meant to protect, was caught in a loop of overreaction.

Then he noticed that his breathing had become subtly more difficult over the past few years. It wasn't full-blown asthma, at least not officially, but in the spring, he often found himself reaching for air when walking through a park or needing to open a window when indoors. His nose was almost always congested, especially in the morning. He had accepted this as seasonal allergies, until he realised they were no longer seasonal. This chronic inflammation of the nasal passages was crowded with eosinophils – white blood cells that acted like overenthusiastic security guards. Whenever James encountered harmless things like pollen or dust, these 'guards' sounded a false alarm, causing swelling, mucus buildup, and congestion. He learned that a particular chemical messenger, called IL-13, made this mucus response even worse, leaving his lungs and nasal passages constantly reacting as if they were under attack – even though there was no real threat.

Food became a complicated subject. Where once he had eaten any food freely, meals now left him bloated, fatigued, or mentally sluggish. Some foods triggered immediate reactions – itchy lips, flushed cheeks, or a tightening in his throat. Others produced more subtle responses: low energy, a heaviness in the gut, or a lingering fog behind his eyes. These were not coincidences. His immune system, primed by Th2 cytokines, had begun generating inappropriate IgE responses to dietary proteins, tagging them as foreign. He was not simply intolerant to food – he was immunologically misinformed.

This misunderstanding extended deep into his gastrointestinal tract. The bloating, loose stools, and cramping were signs of mast cell activation and eosinophilic inflammation that were disrupting normal digestion. He would come across the phrase 'leaky gut' and now understood its immunological basis: Th2 cytokines could disrupt the tight junction proteins – specialised structures that form a protective barrier between cells, preventing pathogens and undigested food particles from entering the bloodstream – in his gut lining. When these tight junctions were compromised, the gut lining became more permeable, allowing food antigens and microbial fragments to pass into his bloodstream unchecked. His immune system responded predictably, triggering more inflammation. And so, the vicious loop continued.

Fatigue became a near-constant companion – not the kind of tiredness that is resolved with rest, but something deeper, more systemic. He often woke up feeling unrefreshed, and by late afternoon, he would be exhausted despite having coffee or any attempts at exercise. He came to understand that inflammation was draining his energy, forcing his body to expend resources to fight threats that weren't truly there. The real culprit, he learned, was a slowdown in mitochondrial function – the internal process by which cells produce energy – caused by his Th2-dominant immune system.

As his energy declined, so too did his cognitive sharpness. James used to pride himself on his sharp intellect and exceptional memory, but he began losing track of sentences, forgetting names, and rereading paragraphs without retention. Brain fog was not a metaphor – it was a lived state. Inflammatory molecules, including histamine and the cytokine IL-4, had crossed the blood-brain barrier and began altering his neurotransmitter landscape. James realised it was not a failure of discipline or related to age; it was immunological dysregulation that spilled into his brain.

Even his hormones seemed to be echoing the imbalance. James had never thought much about oestrogen or progesterone – terms he had associated more with his partner's cycle than with himself. But now, through his learning, he understood how oestrogen dominance – closely linked to Th2 immune activation – could affect men as well. Fatigue, anxiety, sleep disturbances, and even the low libido he hesitated to name out loud, were often the result of hormonal imbalance exacerbated by immune signalling gone awry. His mast cells weren't just releasing histamine – they were also contributing to the inflammatory microclimate that distorted his body's hormonal ratios.

The most difficult part was that none of these symptoms existed in isolation. Together, they formed a web. His foggy brain was linked to his gut; his eczema was connected to his lungs; and his hormones were responding to the same cytokines that influenced his digestion. The Th2 skew wasn't a single disorder – it was a systemic tilt, affecting every surface, every barrier, and every interface between James and his environment.

And yet, even in its complexity, the pattern brought clarity. For the first time, James felt like he had a map – not just of his symptoms, but of the deeper processes driving them. Although his body was exhausted, it was also brilliant and adaptive. It had been working overtime to protect him from threats that no longer needed defending. The overactivation of his Th2 pathways had been shaped by multiple factors – his genetics, ongoing stress, environmental triggers, early immune exposures, a diet high in processed foods lacking essential nutrients, and a sedentary lifestyle that limited the immune-balancing effects of regular physical activity.

What he once saw as unrelated complaints, he now saw as messages – each one a piece of a larger story. And understanding that story would finally liberate him and become his path to healing.

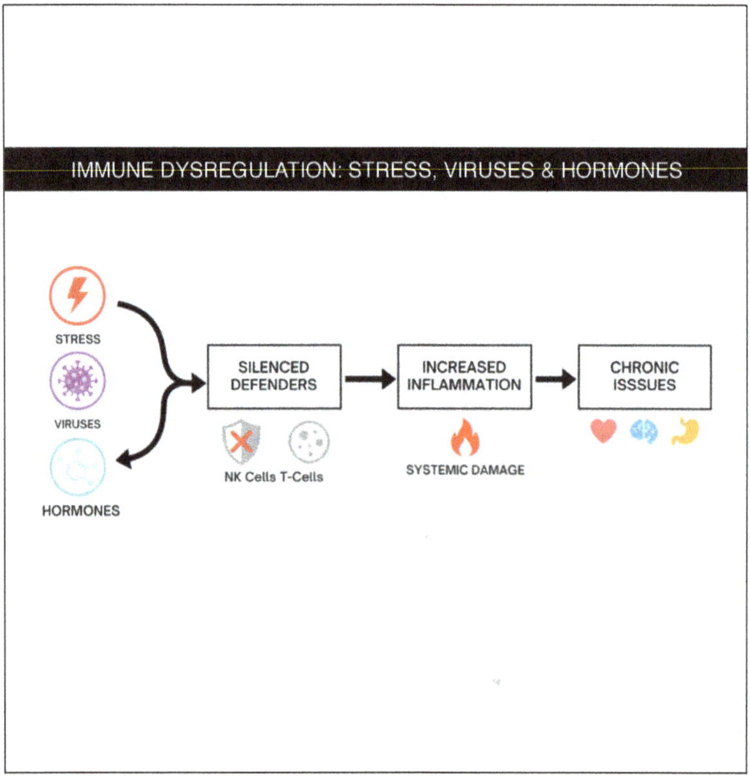
IMMUNE DYSREGULATION: STRESS, VIRUSES & HORMONES
STRESS
VIRUSES
HORMONES
SILENCED DEFENDERS
NK Cells T-Cells
INCREASED INFLAMMATION
SYSTEMIC DAMAGE
CHRONIC ISSSUES

Chapter 9

The Immune Imbalance: What Silences Defenders and Fuels the Fire

James began to understand that his immune system was not simply overactive. It was selectively imbalanced – some parts quieted, others amplified, like a soundboard turned unevenly by years of subtle exposures, chronic tension, and internal misdirection. As he continued his exploration, the picture of Th2 dominance came into sharper focus. It was no longer only about allergies or inflammation, but a *disrupted immune hierarchy* – where his innate surveillance had gone quiet, and a hypersensitive, allergic bias had taken over.

He read that his *NK cells*, key components of innate immunity, were meant to serve as first responders – quietly identifying and eliminating virally infected or abnormal cells before they could take root. Yet, under chronic stress their function was heavily compromised. Years of stress had kept his cortisol levels elevated, slowly but steadily *dampening NK cell activity*. Studies confirmed that prolonged cortisol exposure suppressed NK cell cytotoxicity and proliferation (Segerstrom and Miller, 2004). James had felt the shift: viral infections that lingered, a susceptibility to every seasonal bug, and a sense that his body was no longer mounting defences effectively.

But cortisol didn't just quiet NK cells. It was also a messenger – shifting the immune tone in a very specific direction. Under its influence, the body began releasing *IL-4 and IL-5*, hallmark Th2 cytokines. These weren't inherently harmful; they were part of the body's design to handle parasites and wound repair. But in James' case, they had become dominant, steering the immune system into hypersensitivity. His hay fever, eczema, and histamine reactions all made sense now – not as separate issues, but as part of a pattern where *stress had downregulated his defenders and upregulated his reactivity* (Rohleder, 2014).

There was another layer he couldn't ignore. As mentioned earlier in this book, James had tested positive for *Epstein-Barr virus.* He thought it was a passing infection, but now he realised EBV doesn't always leave – it lingers in the background, adapting and manipulating the host immune system to ensure its survival. EBV, like other chronic viruses, actively *suppresses NK cells*, exhausting them over time through persistent antigenic stimulation. More disturbingly, it secretes immune-modulating proteins like *viral IL-10*, tipping the immune response towards a Th2 profile to avoid being destroyed (Jochum et al., 2012). The virus's strategy was clever: weaken the immune surveillance that would eliminate it and amplify the Th2 responses that would ensure its persistence.

These internal shifts were amplified by changes at the molecular level. James learned that regulatory cytokines such as *IL-10* and *TGF-β*, usually tasked with keeping immune responses in check, could also be hijacked to further imbalance. In excess, IL-10 suppressed NK cell development and reduced their ability to produce interferon-gamma. TGF-β, a molecule often elevated during chronic immune activation, directly blocked NK cytotoxicity. At the same time, these same cytokines helped promote *Th2 differentiation*, driving the immune system to release more IL-4, IL-5, and IL-13 – all inflammatory mediators now found in excess in his lungs, skin, and gut (Akdis et al., 2020). These were not just responses to allergens. They were systemic consequences of long-standing immunological redirection.

Hormonal factors added yet another dimension. For years, James hadn't thought of oestrogen as relevant to his own biology, but it became clear that this hormone played a powerful role in immune regulation across sexes. Elevated oestrogen levels, whether from excess adipose tissue or endocrine-disrupting exposures (such as chemical pollutants), were known to *inhibit NK cell cytotoxicity*. At the same time, oestrogen had a profound influence on T cell differentiation, where immature T cells specialise into functional T cells, promoting the survival and activity of *Th2 cells*. This relationship explained why women are more susceptible to Th2-driven conditions like asthma and eczema, particularly during periods of hormonal fluctuation such as menstruation or pregnancy (Taneja, 2018). For James, it contextualised his own subtle hormone imbalance and the hormonal imprint of stress and diet on his immune system.

But what about the world James lived in? The air he breathed, the food he ate, and the soil he walked on – were also shaping his immunity. Environmental toxins, particularly persistent organic pollutants like dioxins and polychlorinated biphenyls, had been shown to *weaken NK cell function* through oxidative stress and mitochondrial damage. They also influenced immune regulation by binding to the *aryl hydrocarbon receptor (AhR)* – a molecular switch that, when triggered, promoted *Th2 cytokine expression* and allergic inflammation (Kerkvliet, 2002). For James, the daily exposures that once seemed harmless – car fumes, cleaning sprays, even traces of pesticides in food – were now seen as subtle reprogrammers of his immune architecture.

What surprised him most, perhaps, was the impact of simple nutritional insufficiencies. James had always considered himself fairly healthy, even though his diet had been inconsistent, especially during stressful periods. He now realised that micronutrients like *vitamin D* and *zinc* were not just supporting players – they were *essential cofactors for NK cell maturation and function*. Without them, his cytotoxic defences weakened, and his immune responses shifted further towards Th2 dominance. Vitamin D deficiency had been shown to reduce interferon-gamma production, impair NK cell and Th1 responses, and foster allergic immune patterns. Zinc, meanwhile, supported NK cell enzyme systems and helped modulate T cell signalling (Gombart et al., 2020).

As James put all these pieces together, he stopped thinking of his immune imbalance as something random or genetically inevitable. It had been shaped by stress, infection, hormones, toxins, and micronutrient gaps. Each of these influences hadn't just 'weakened' his immunity; they had *redirected it*, silencing the cells that defended him from chronic infections and malignancy, and *elevating the ones that inflamed him in response to dust, dander, and dinner*.

However, James now saw a cohesive map. His immune system had been trying to adapt, to survive, and to protect. But it had learned the wrong lessons. His NK cells had gone quiet in the face of relentless pressure, while his Th2 cells had grown louder, more reactive, and more prone to seeing the world as a threat. This wasn't just autoimmunity. It wasn't just allergy. It was a *misguided survival strategy* – one that had turned inward, making health feel like a battle.

But maps, he reminded himself, are not meant to point out the problem – they are meant to show the way forward. And now that he could see where his immune system had wandered, he could begin the long and deliberate work of bringing it back into balance.

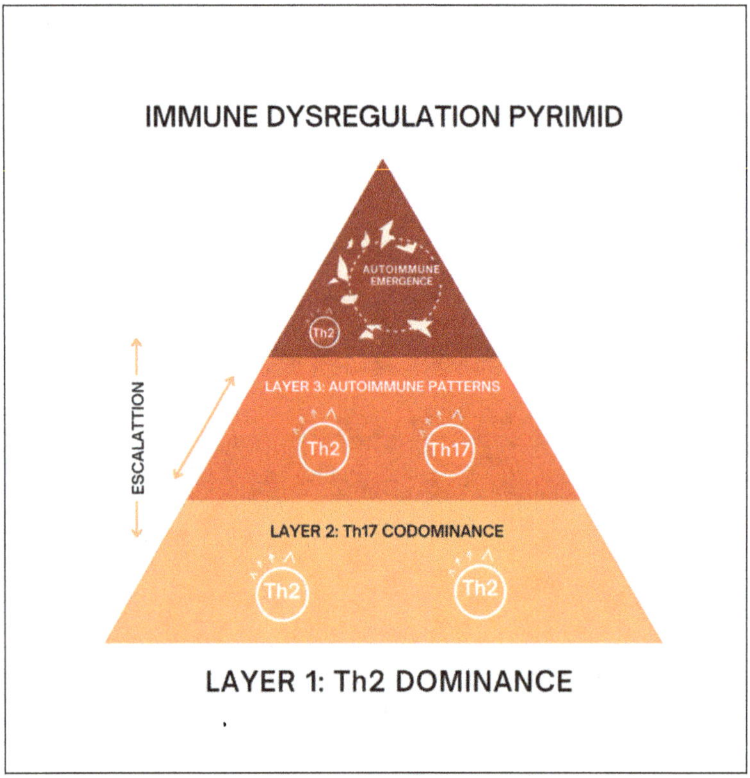
IMMUNE DYSREGULATION PYRIMID
AUTOIMMUNE EMERGENCE
Th2
LAYER 3: AUTOIMMUNE PATTERNS
Th2
Th17
ESCALATTION
LAYER 2: Th17 CODOMINANCE
Th2
Th2
LAYER 1: Th2 DOMINANCE

Chapter 10

The Tipping Point: When Inflammation Layers Itself

For a while, James believed that if he could quiet his allergies, calm his skin, and soothe his gut, he would have restored balance. But the further he followed the path of immune dysregulation, the more he came to understand that his body was not merely inflamed – it was being reprogrammed.

What had started as Th2 dominance – an immune tilt towards allergic, histamine-fuelled reactivity – had become something more layered, more embedded. There were other players now: *iTregs*, *Th17 cells*, and *neutrophils*, all entangled in a deeper choreography of dysregulation. James had once felt like his immune system was overreacting to the outside world. Now he could see how it was beginning to *overreact to itself.*

He learned that *induced regulatory T cells (iTregs)* were the peacemakers of the immune system – cells designed to keep inflammation in check, to prevent overreactions and autoimmune loops. But in Th2-dominant environments, these peacemakers were being silenced. The cytokines that flooded his system – *IL-4* and *IL-13* – not only drove allergic inflammation but *inhibited the development of iTregs*, reducing the levels of IL-10, the very molecule that acted like a brake on runaway immunity (Chapoval et al., 2010; Schmitt and Williams, 2013). Without IL-10, his immune system lost one of its most powerful tools of restraint.

This loss created space for another lineage to rise. *IL-6*, often elevated in inflammatory environments, was no longer being held in check. It began to synergise with *TGF-β*, not to produce iTregs as it might in a regulated system, but to signal instead for the development of *Th17 cells* – a shift that changed the terrain entirely.

These Th17 cells were not subtle. They released *IL-17*, *IL-21*, and *IL-22* – cytokines known for their roles in recruiting *neutrophils*, amplifying inflammation, and damaging tissue over time (Hagihara et al., 2019). In small, acute doses, these cytokines could help the body clear infections and repair tissue. But in James' case, the Th17 response had become chronic. His gut lining – once a resilient barrier – now oscillated between attempts at repair and episodes of deepened permeability. *IL-22*, which in health supports epithelial (gut lining) repair, had become a double-edged sword – *contributing to barrier dysfunction* and inflammatory relapse (Schmitt and Williams, 2013).

But perhaps the most tangible sign of this shift was what James now felt during a flare. There was a heat to it – not just in his skin, but in his chest, his joints, even his thoughts. This was the work of *neutrophils*, the immune cells called in by IL-17, now pouring into tissues and *releasing enzymes and reactive oxygen species*. Once tasked with defending him against infection, these cells were now *digesting the fabric of his own tissues*, triggering *fibrin deposition* (which is explained in the next chapter), the thickening of mucous layers, and oxidative cascades that sustained inflammation long after the initial trigger had passed.

He saw now how inflammation was no longer a single stream but a *layered network*, where *Th2 and Th17 activity overlapped*, feeding off each other. While these pathways often counter-regulate in acute immune responses, James' immune system had lost that elegant balance. Th2 dominance had weakened iTreg formation and allowed IL-6 to dominate, driving Th17 responses. These arms of immunity – meant to be distinct – were now working in tandem, creating a *perpetual inflammatory loop* (Chapoval et al., 2010).

James also noted how this inflammatory synergy showed up systemically. On bad days, his skin flared, his digestion unravelled, and his cognition slipped – not all at once, but in sequence, like a domino effect. His body was no longer reacting to allergens alone. It was responding to *its own unmet regulation* – to the internal chaos that had replaced the immune order he never knew he needed to protect.

This new understanding brought a strange sense of relief. These processes weren't random, nor meaningless. They were the logical outcome of a system trying to adapt and maintain equilibrium in the absence of regulatory leadership.

James now understood that his Th2 dominance was not just an allergy problem – it was the *doorway to deeper dysfunction*. The quieting of iTregs, the rise of Th17 inflammation, and the activation of neutrophils – all of it was part of the same immune choreography, one that had begun with subtle skewing and was now *shaping the very terrain of his chronic inflammation*.

It was no longer about symptom management. It was about *rewriting the instruction manual* – understanding what suppressed iTregs and what nurtured them, what fuelled IL-6 and what helped quiet it, what signalled his immune system towards chaos and what could pull it back towards coherence.

For the first time, James felt ready not just to treat what was happening on the surface, but to address the layers beneath – *to understand the architecture of inflammation*, not merely its flames.

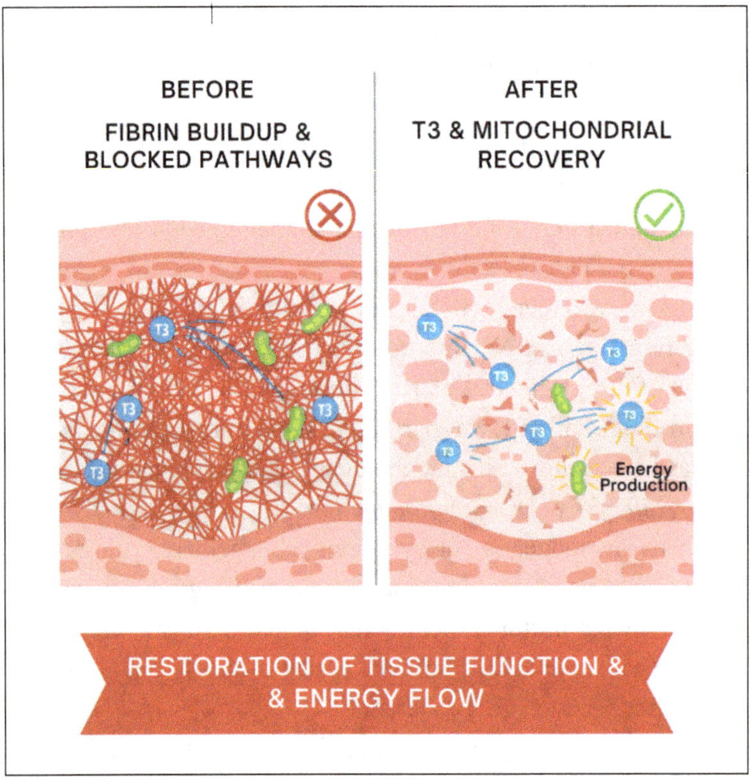
BEFORE
FIBRIN BUILDUP &
BLOCKED PATHWAYS
T3
AFTER
T3 & MITOCHONDRIAL
RECOVERY
T3
Energy
Production
RESTORATION OF TISSUE FUNCTION &
& ENERGY FLOW

Chapter 11

The Slowing Fire: Fibrin and the T3–Mitochondrial Disconnect

James had grown familiar with the sensation of fatigue – the kind that could not be resolved with rest. At first, it was easy to blame it on stress, poor sleep, or even emotional weight. But now, having travelled so far into the labyrinth of his immune system, he sensed something deeper. His exhaustion, caused by inflammation, was also the result of energy – or more precisely, his *body's growing inability to produce it efficiently*.

He began to explore the machinery of his cells: the mitochondria. Once known to him only as 'the powerhouse', he now saw them as sensitive, finely-tuned command centres, deeply influenced by *thyroid hormones*, especially *T3 (triiodothyronine)*. T3, he learned, didn't just influence metabolism from the top down – it was a direct messenger to mitochondria, telling them when to burn fat, when to metabolise glucose, when to generate heat, and when to rest. But in his case, those messages were not getting through.

The missing piece was *fibrin* – a substance he had only heard of in the context of clotting. But now he discovered its role far beyond coagulation. In chronic inflammatory states, fibrin wasn't just forming clots; it was silently *accumulating in his tissues*, thickening the microvasculature, altering the very structure through which oxygen, nutrients, and hormones travelled.

It made sense. If *T3 needed to reach his cells*, it had to pass through the blood vessels, cross the endothelium (the layer of cells lining the blood vessels), and enter the mitochondria (the inner 'engine' of the cell). But where *fibrin built up*, it *thickened capillary walls*, *triggered endothelial dysfunction*, and *narrowed vascular flow*. As a result, his T3 was not vanishing – it was getting stuck, diluted in transit, unable to reach the very tissues that needed it most (Squizzato et al., 2007).

His vascular system was no longer a clear passage – it had become clogged with the residue of chronic inflammation. The endothelium, once flexible and responsive, was now stiffened by oxidative stress, its tight junctions (the protective seal between cells) impaired. As fibrin built up, it became a *scaffold for immune cells*, amplifying local inflammation and *triggering the release of cytokines and ROS.* These inflammatory signals further *damaged the endothelial lining*, creating a barrier not only to oxygen but to *thyroid hormone transport* itself (Luyendyk et al., 2019).

He had always assumed that his thyroid function was 'normal' because his TSH (thyroid-stimulating hormone) hadn't raised alarms. But now, he saw how function could be disrupted *not at the level of hormone production, but at the level of delivery and response.* Even if T3 was present in the bloodstream, it was not reaching its intracellular targets. And even when it did, the mitochondrial *receptors were compromised* – oxidised, inflamed, and unresponsive (Pingitore et al., 2016).

This was not hypothyroidism in the traditional sense – it was *tissue-level hypothyroidism*, one driven not by the thyroid gland but by *vascular obstruction, mitochondrial resistance, and immune dysfunction*.

As he read further, James learned how essential T3 was to the very processes he struggled with: *lipid (fats) oxidation*, *glucose metabolism*, and *thermogenesis (body temperature regulation).* T3 modulated the activity of key mitochondrial enzymes that governed metabolic flexibility – his ability to shift between burning fat and glucose, depending on demand. Without T3, that flexibility was lost. His cells became inefficient, sluggish, accumulating lipids they could no longer burn and glucose they couldn't fully utilise. He saw this not just in blood tests but in his lived experience – *weight that wouldn't shift, crashes after meals, a body that felt cold and slow despite adequate nutrition* (Fröhlich and Wahl, 2021).

Fibrin, he realised, had quietly woven itself into every aspect of his bodily dysfunction. It was interfering with T3 transport, yes – but it was also indirectly *shaping his metabolism*, his energy, and even his cardiovascular risk. The more he learned, the more he understood how *fibrin formed a bridge between immune dysfunction and endocrine disruption.* Where there was unresolved inflammation, there was fibrin; where there was fibrin, there was endothelial stiffness; and where T3 couldn't reach, *mitochondria fell silent.*

And so, the fatigue he felt was not just biochemical – it was architectural. His internal landscape had changed. Systems designed to be responsive were now rigid. Signals meant to be dynamic were now muffled. What he once interpreted as tiredness was, in fact, the *expression of blocked metabolic communication* – his body's fire slowly dampened, not from lack of fuel, but from the inability to respond to it.

But if dysfunction had built up over time, James believed, so too could healing. What if fibrin could be broken down? What if endothelial function could be restored? What if mitochondria could regain their sensitivity to T3? These were no longer abstract questions – they were invitations.

He began to imagine what it might mean to not just restore thyroid levels on paper, but to *restore thyroid responsiveness in body tissue* – to revive his metabolic rhythm by clearing the static between messenger and receiver.

James' story was no longer about suppressing inflammation or managing symptoms. It was about *clearing the path* – for hormones, for energy, for breath, and for life to move freely again.

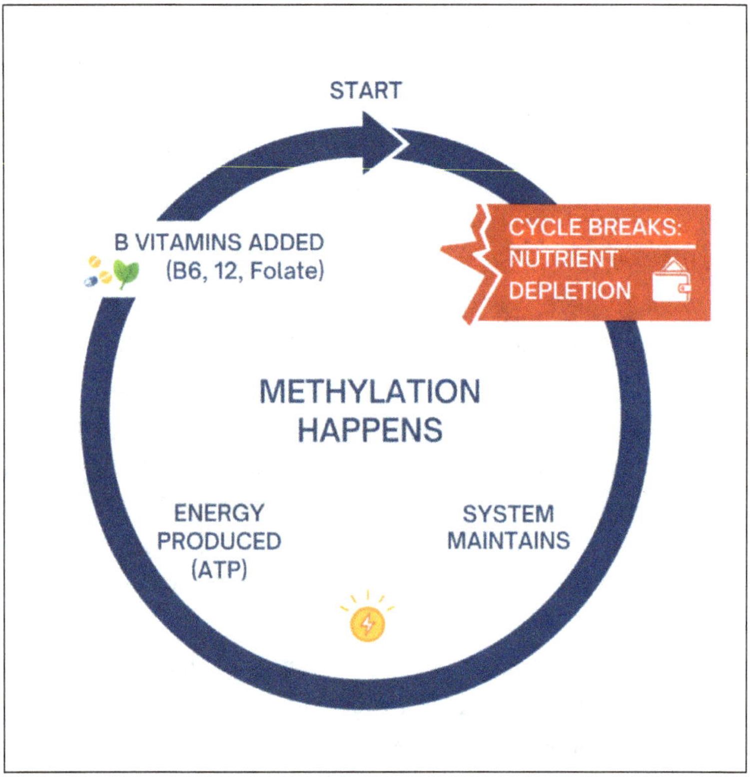
START
B VITAMINS ADDED
(B6, 12, Folate)
CYCLE BREAKS:
NUTRIENT
DEPLETION
METHYLATION
HAPPENS
ENERGY
PRODUCED
(ATP)
SYSTEM
MAINTAINS

Chapter 12

The Quiet Collapse: Methylation Depletion in Th2 Dominance

James had grown used to thinking about inflammation as loud – rashes, flares, wheezing, and congestion. But now he was discovering another side to dysfunction. One that was quieter, subtler, and in many ways, more insidious.

In chapter 10 the layers of inflammation are discussed. But this chapter is about the body's *inability to clear the inflammation.* His detoxification systems, neurotransmitters, and antioxidant reserves were not only under pressure – they were *starved.* The common thread tying them together was methylation.

This system, once invisible to him, was now at the centre of everything. Methylation, he learned, was more than a buzzword – it was a *foundational biological rhythm*, responsible for turning genes on and off, regulating neurotransmitters, generating glutathione (an antioxidant produced in the cells), and detoxifying hormones and toxins. And it was deeply impaired in people like James – those burdened by *Th2-dominant immune profiles*, chronic oxidative stress, and ongoing inflammation.

He imagined methylation like a currency. Every time his body needed to neutralise a toxin, synthesise serotonin, recycle homocysteine, or produce glutathione, it spent a molecule of that currency – often in the form of *S-adenosylmethionine (SAMe),* which helps to regulate enzymes that influence communicative molecules such as hormones and neurotransmitters.

And like any overdrawn account, it wasn't long before the system began to default.

Without sufficient *SAMe*, the neurotransmitter balance began to falter. He had noticed it as emotional volatility – on days where he couldn't explain his irritability or sense of depletion. Behind the scenes, low SAMe and low *tetrahydrobiopterin (BH4)* (essential for metabolic processes, particularly neurotransmitter synthesis) were slowing the synthesis of neurotransmitters *dopamine, serotonin, and norepinephrine* (Kapatos, 2013). What once felt like burnout was, at least in part, biochemical – his mood blunted by a system unable to keep up with demand.

The same held true for his resilience to stress. Without methyl donors (the molecules that act as switches to gene expression) and with BH4 depletion, the enzyme *nitric oxide synthase (NOS)* became uncoupled, producing *superoxide* instead of *nitric oxide (NO)*. The consequences were vascular as well as emotional – *impaired blood flow, brain fog, and a nervous system wired for agitation rather than rest* (Alp and Channon, 2004).

And then there was *glutathione*, the master antioxidant. For James, this molecule had always felt like a buzzword until he understood what happened without it. In the context of *Th2-driven inflammation*, *folate depletion*, and *SAMe loss*, glutathione synthesis faltered (Lu, 2013). What was left behind was oxidative residue – damaged proteins, misfiring enzymes, and a system on the edge of collapse.

He now understood that methylation and detoxification were not separate – *they were the same conversation*. His *SUOX enzyme*, tasked with converting *sulphite to sulphate*, was underperforming. And in its dysfunction, *sulphite accumulated*, triggering further oxidative stress and inflammation. Worse, this toxic buildup drained the very glutathione reserves he needed to protect his mitochondria and clear toxins (Mendel and Schwarz, 2023).

A similar story unfolded in the *SULT* pathways – enzymes that required sulphate to deactivate and excrete hormones like *oestrogens* and chemicals like *xenobiotics*, which are foreign molecules that mimic innate particles – overriding biological function and dysregulating physiology. But under Th2 dominance, methylation loss and sulphur depletion made it impossible for these systems to keep up. Unmetabolised oestrogens accumulated, driving inflammation. Toxins remained in circulation. His detox pathways were congested – not because they were lazy, but because they were *undernourished and overburdened* (Glatt, 2000; Bottiglieri, 2002).

Another underappreciated casualty of SAMe depletion was the *polyamine pathway*. Normally, SAMe fuels the production of *polyamines* like *spermidine*, which support *neural repair, immune balance, and cellular regeneration*. But with SAMe diverted or depleted, polyamine synthesis plummeted. The result was not only a slowing of cognitive and immune repair – it was an *erosion of cellular adaptability* itself (Pegg, 2009).

Every system James had come to rely on – his energy, his mental clarity, his emotional balance, and his immune regulation – was being quietly degraded. Not by one massive inflammatory flare, but by a *steady depletion of methylation capacity*. The deeper his Th2 dominance became, the more oxidative stress was generated. The more oxidative stress he endured, the more methylation demand rose. And the more his methylation pathways struggled, the harder it became to generate SAMe, BH4, folate, and glutathione – the very molecules needed to bring balance back.

It was a *feedback loop with no natural off switch*.

But seeing the pattern changed everything. James was no longer fighting isolated symptoms – he was looking at a system that was deprived of the raw materials it needed to regulate itself. His healing wasn't just about suppressing inflammation or neutralising histamine – it was about *restoring capacity* by giving his body the tools to methylate, detoxify, and recover.

And so, his next steps would be different. Instead of chasing symptoms, he would begin *rebuilding the currency* of health by supporting methylation with nutrients such as B12, folate, and betaine; restoring BH4 through antioxidants and nitric oxide balance; and unburdening SUOX and SULT systems with targeted cofactors and gentle detoxification. This was no longer a battle – it was an investment.

An investment in coherence. In connection. In the restoration of a rhythm that his body had been quietly begging to remember.

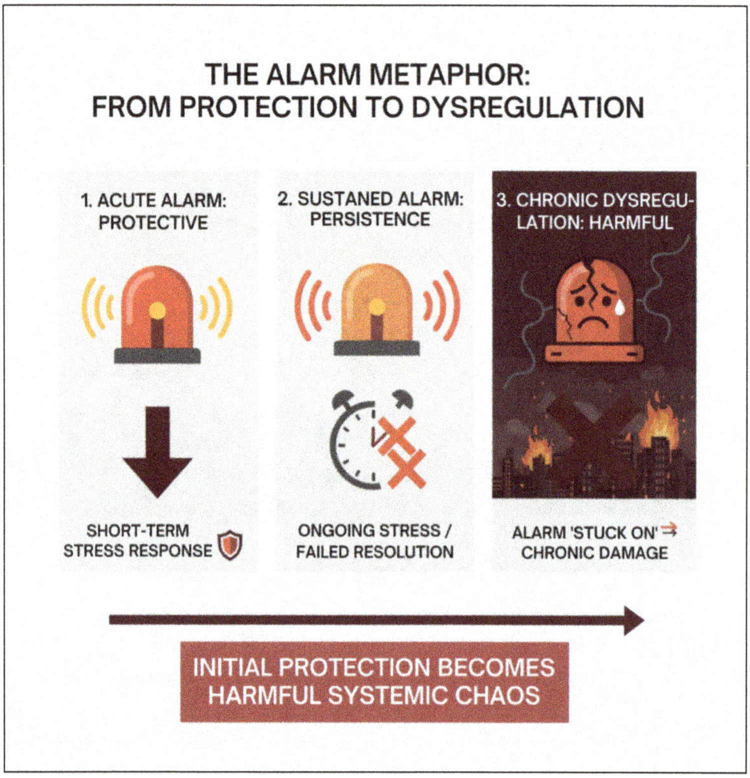
THE ALARM METAPHOR:
FROM PROTECTION TO DYSREGULATION
1. ACUTE ALARM: PROTECTIVE
2. SUSTANED ALARM: PERSISTENCE
3. CHRONIC DYSREGU- LATION: HARMFUL
SHORT-TERM STRESS RESPONSE
ONGOING STRESS / FAILED RESOLUTION
ALARM 'STUCK ON' → CHRONIC DAMAGE
INITIAL PROTECTION BECOMES HARMFUL SYSTEMIC CHAOS

Chapter 13

The Alarm That Would Not Turn Off: Th2 Dominance and the Cell Danger Response

There was a point in James' journey when it all began to feel connected – not just in a conceptual sense, but biologically. His fatigue, brain fog, immune instability, hormonal imbalances, and gut disruption were not separate events; they were chapters of the same story. And as he looked deeper, he found its title: the *Cell Danger Response (CDR).*

Dr Naviaux's work on the *CDR1* state was a revelation. James had assumed his immune system was reacting randomly – hypersensitive to the benign, sluggish towards the harmful. But now, he could see how his body had entered a state of *cellular defence* – a metabolic lockdown in response to unresolved threats. The problem was that *the alarm never shut off* (Naviaux, 2014).

Th2 dominance had created the ideal environment for this chronic defence posture. On one side, his immune system was reactive – producing IL-4, IL-5, and IL-13 cytokines, which induce Th2 responses such as increased histamine release, in response to harmless stimuli. On the other, it was *failing to mount effective surveillance*, leaving behind low-grade infections, oxidised molecules, and dysfunctional mitochondria. This wasn't just hyperactivity – it was immune confusion, causing further problems such as increased inflammation, fatigue, allergies, and difficulty fighting infections.

James often worried about his mental clarity – or lack thereof. He had read that *Th2 dominance suppressed Th1 responses*, which were critical for neuroprotection. Without this balance, the brain lost its ability to clear *beta-amyloid plaques*, and inflammation smouldered unchecked. In the presence of mitochondrial dysfunction, oxidised *mtDNA* (mitochondrial DNA – genes that regulate the energy of the cells' powerhouse) spilled into the extracellular space, activating the *NLRP3 inflammasome* – a fire alarm for the immune system. But instead of mobilising efficient repair, it triggered a neuroinflammatory loop, further entrenching *synaptic degradation and beta-amyloid accumulation.*

The fog he experienced was not simply cognitive – it was *mitochondrial and immune in origin.*

Then came the realisation that his long-standing fatigue, the kind no sleep could touch, was more than hormonal imbalance or nutritional deficit. It had a viral fingerprint. In his earlier years, James had contracted the Epstein-Barr virus, causing a lingering fatigue ever since. He now knew that *Th2 dominance* and its suppression of *NK cells and Th1 activity* created a blind spot for the immune system – allowing intracellular pathogens like EBV and HHV-6 to *persist.* Their presence not only activated immune pathways continuously but also *fragmented mitochondria*, pushing cells deeper into a hypometabolic, energy-conserving mode – a classic feature of the *CDR1 state* (Schreiner et al., 2020).

His muscles now fatigued quickly, and his brain lost its sharpness. He now understood that *CDR1 inhibits mitochondrial oxidative phosphorylation* (one of three respiratory systems of the mitochondria producing energy, responsible for producing 90 per cent of total energy yield), slowing ATP production (the cell's energy) and increasing *ROS generation*, which, in excess, can damage cells. Mitochondrial membranes became damaged, and the receptors desensitised. His cells weren't just tired – they were *actively disengaged* from full metabolic function in order to survive.

This shutdown was not limited to his brain and muscles. His immune system was changing shape. *Tregs, already suppressed in Th2 dominance*, were further diminished by the presence of oxidised mtDNA – classified by the body as *'Damage-Associated Molecular Patterns'* (*DAMPs*). These signals kept innate immune cells in a state of activation, while breaking the feedback loops that normally restored tolerance. What emerged instead was *autoimmune potential* – an immune system so dysregulated that it began targeting itself.

James recognised the pattern: *lupus-like rashes*, *joint stiffness*, and *seasonal flare-ups* that once seemed random were, in fact, expressions of an immune system caught in a loop – Th2-driven on the surface, but underneath, *locked in a danger posture*, perceiving even the body's own tissues as threats (Zolkipli-Cunningham et al., 2021).

The final blow came in the form of quiet fear – one that many chronically inflamed bodies share, even if unspoken: *the risk of cancer*. James learned that *IL-4 and IL-13*, while part of the Th2 repertoire, also help tumours thrive. They *suppress cytotoxic T cells and NK activity*, creating an immunosuppressive tumour microenvironment. In parallel, the CDR1-induced metabolic shift supports *aerobic glycolysis* – the so-called *Warburg effect* – which allows cancer cells to thrive in hypoxic (low levels of oxygen), inflamed conditions (Naviaux et al., 2017).

His body was not just misfiring – it was being hijacked. Not by a single enemy, but by a series of *compensatory mechanisms that had become maladaptive*. Th2 dominance and the CDR1 state were two sides of the same coin – one inflaming from the outside, the other constraining from within the cells. Together, they created a terrain of *low energy, high inflammation, impaired repair, and immunosuppression*; a terrain that breeds persistence, not resolution.

Yet within this clarity, James felt something unexpected: empowerment.

Because these were not irreversible patterns. They were *adaptive responses* to real, unresolved threats – trauma, infection, environmental burden, and emotional pain. His body wasn't defeated. It was still trying to protect him, even now. The challenge ahead was not to suppress these systems, but to *gently signal safety* – to support the mitochondria, calm the immune system, and resolve the stored danger.

The answer, he knew, would involve good nutrition, as well as movement, nervous system recalibration, mitochondrial cofactors, antiviral strategies, and emotional repair. It would, however, take time. And now he had something even more important than energy. He had a map.

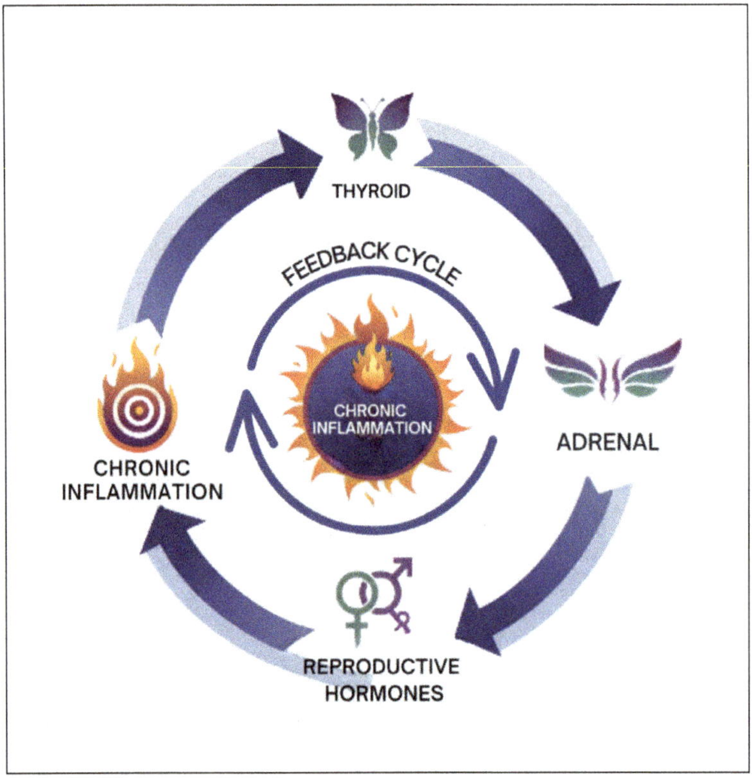

THYROID
FEEDBACK CYCLE
CHRONIC
INFLAMMATION
ADRENAL
CHRONIC
INFLAMMATION
REPRODUCTIVE
HORMONES

Chapter 14

The Endocrine Echo: When Hormones Follow Inflammation

James had always understood hormones in fragments – cortisol during stress, thyroid for energy, melatonin for sleep. They seemed to operate in isolation, each with its own rhythm and logic. But as his investigation deepened, he came to see that hormones were not independent signals. They were reflections, or echoes, of deeper patterns. And now, those patterns were starting to make sense.

Not only was his immune system dysregulated, so too was his entire *neuroendocrine architecture*. The long-standing dominance of Th2-driven inflammation had begun to shape his *hypothalamic-pituitary-adrenal (HPA)* and *hypothalamic-pituitary-thyroid (HPT)* axes in ways that felt subtle at times, yet were systemic and pervasive.

His days no longer followed a predictable arc of energy and rest. Instead, he felt wired when he needed to sleep, and flat when he needed to move. He was waking in the early hours of the morning with a racing heart, only to crash mid-afternoon with fatigue so heavy that it blurred his thoughts. What he once thought was poor sleep hygiene, he now recognised as *HPA axis dysfunction*.

At the root of it was *inflammation*. The cytokines dominating his system – *IL-4*, *IL-13*, and *IL-10* – were interfering with his hypothalamus, disrupting the release of *corticotropin-releasing hormone (CRH)*, which is supposed to regulate his circadian cortisol rhythms.

Instead, CRH secretion was misaligned, leading to *excessive HPA activation*, with cortisol levels that fluctuated inappropriately – often *too high at night and too low in the morning.*

But cortisol was no longer doing its job effectively. Chronic stress and prolonged cortisol elevation had led to *glucocorticoid receptor resistance*, a condition where the cells that once responded to cortisol's anti-inflammatory signals were now numb. This insensitivity meant that inflammation no longer shut off. It also meant that cortisol kept rising, driving *further suppression of Th1 immunity*, and locking James deeper into *Th2 dominance* (Schreiner et al., 2020).

He was trapped in a *loop* – a feedback cycle in which inflammation increased cortisol, and cortisol prolonged inflammation. And when glucocorticoids lost their signalling power, the *Cell Danger Response* could not resolve it. His body remained in a state of low-grade alert – drained, edgy, and inflamed.

But the HPA axis wasn't the only system that was disrupted.

His *thyroid*, too, had been showing signs of distress. For years, his lab test results were 'normal', but he had felt far from well. He was cold when others were not, his weight crept up even as he watched his diet, and his muscles ached after mild exertion. Now he understood that *Th2 cytokines* and inflammatory mediators like *IL-6* and *TNF-α* were interfering with both *thyrotropin-releasing hormone (TRH) secretion from the hypothalamus* and *thyroid-stimulating hormone (TSH) responsiveness in the pituitary gland.* More critically, they were *inhibiting the activity of deiodinase enzymes* – the very tools his body used to convert thyroid hormones *thyroxine (T4) into active triiodothyronine (T3)* (Schreiner et al., 2020).

Even if his TSH was within range, his *free T3 was low*, and his *reverse T3 was climbing* – a picture consistent with the often-overlooked *euthyroid sick syndrome* or *low T3 syndrome*. This was not classical hypothyroidism – it was a *tissue-level thyroid resistance*, compounded by *mitochondrial dysfunction* from CDR1 activation. His cells weren't using thyroid hormone effectively because they were too oxidised, too inflamed, and too stuck in survival mode to respond (Zolkipli-Cunningham et al., 2021).

Mitochondria, once sensitive to T3's metabolic signals, had lost their responsiveness. In this altered state, thermogenesis and lipid metabolism slowed. His body could no longer flex between fuels, and his energy economy tanked. This explained the unrelenting *fatigue*, the *cold hands*, the *slow recovery from stress*, and the weight that refused to shift despite effort. His metabolism had *shut itself down to conserve resources*.

On some days, there were even flickers of the opposite: *surges of anxiety*, heart palpitations, and sudden heat in his chest. These, he now recognised, were *transient spikes in thyroid activity*, a phenomenon seen in *Hashimoto's thyroiditis*, where the immune system attacks the thyroid in cycles – alternating between phases of hormone release and depletion. Th2 dominance, he had learned, *disrupts immune tolerance*, weakening *Treg surveillance*, and increasing the risk for *thyroid autoantibodies*. For James, the storm within had reached his thyroid gland.

These were no longer mere abstract mechanisms – they were *explanations for years of lived experience*. His cortisol wasn't just a stress hormone – it was the product of immune miscommunication. His thyroid wasn't just sluggish – it was *under siege*, both from outside (inflammation) and within (autoimmunity). Both systems were reflecting back the imprint of a deeper imbalance: *Th2 dominance sustained by chronic immune activation and CDR1 signalling*.

But understanding this cascade also meant the possibility of intervention.

If inflammation was distorting hormone rhythms, it could be calmed. If oxidative stress was blocking receptor function, it could be buffered. If receptor resistance was entrenched, it could be retrained – slowly, gently, through metabolic signalling, circadian alignment, micronutrient repletion, and nervous system repair.

This chapter of James' story wasn't just about what had gone wrong – it was about *recognising the pattern*, tracing the thread back to its root, and beginning to *unwind it*.

What looked like burnout, depression, and chronic illness, was in truth *the body's attempt to survive a prolonged alarm* – its refusal to run a full metabolism when the world inside still felt unsafe. The endocrine system wasn't broken. It was brilliant. It had done its job.

Now, it was time to teach it that the threat had passed.

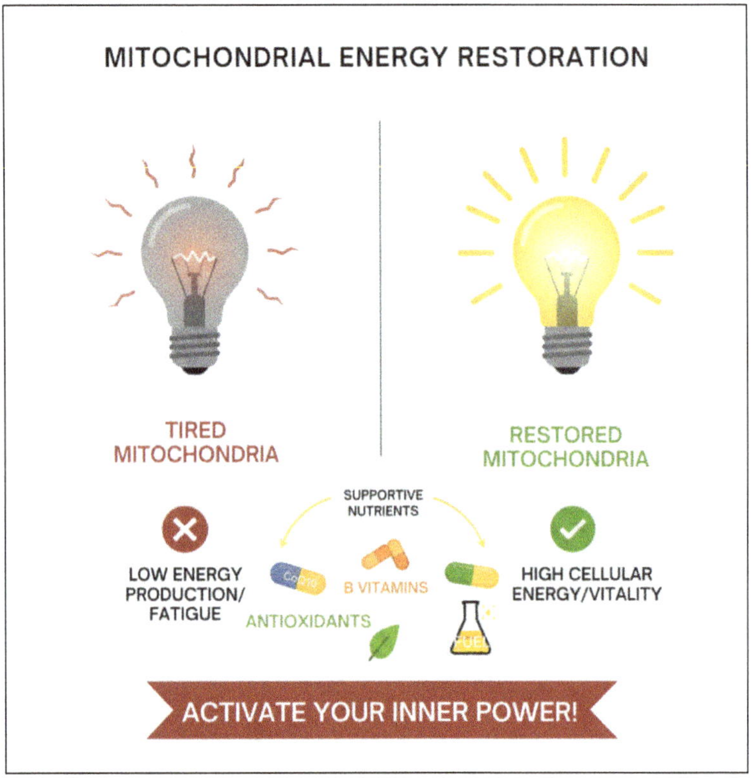
MITOCHONDRIAL ENERGY RESTORATION
TIRED MITOCHONDRIA
RESTORED MITOCHONDRIA
SUPPORTIVE NUTRIENTS
LOW ENERGY PRODUCTION/ FATIGUE
B VITAMINS
ANTIOXIDANTS
HIGH CELLULAR ENERGY/VITALITY
ACTIVATE YOUR INNER POWER!

Chapter 15

Reigniting the Fire: Solutions to Improve Mitochondrial-Thyroid Receptivity

James stood at the edge of something new – not just a deeper understanding of his physiology, but the first glimpse of *recovery as reconstruction*. If Th2 dominance, chronic inflammation, and the Cell Danger Response had eroded the relationship between his mitochondria and thyroid hormones, then his healing would require *rebuilding that connection, molecule by molecule, signal by signal.*

He had spent years trying to optimise thyroid health through routine labs and surface-level interventions. But now he understood that the real breakdown wasn't in the gland – it was in *receptivity*. His cells were no longer responding to thyroid hormones, particularly *T3*, the active form that communicated directly with mitochondria. However, the mitochondria themselves – damaged, oxidised, fragmented – had forgotten how to listen.

The first step was clear: *optimise T3 availability*. James learned that TSH and even T4 could remain 'normal' while *free T3 (fT3)* plummeted. In his case, reverse T3 (rT3) had crept up – blocking receptor access like static noise on a radio channel. Through careful guidance, he transitioned from T4 monotherapy to a *T3-inclusive protocol*, allowing direct stimulation of mitochondrial pathways, particularly in energy-intensive tissues like the brain, heart, and skeletal muscle (Kelly, 2000). This was not about raising levels – it was about *restoring responsiveness*.

But he knew hormones could not work in isolation. Years of *oxidative stress* had degraded the mitochondrial membranes and altered their receptors. He began building an antioxidant shield – not with megadoses, but with precision. *Selenium*, *glutathione*, and *N-acetylcysteine (NAC)* formed the core of his protection, quenching free radicals that impaired thyroid receptor function. He added *zinc*, *copper*, and *manganese* – not for immune health alone, but to support *superoxide dismutase (SOD)*, the mitochondrial enzyme that fought back against ROS from within (Rayman, 2000). As his internal terrain quieted, he turned to the next frontier: *mitochondrial biogenesis*. He learned that damaged mitochondria were not necessarily a dead end – cells could build new ones, if properly signalled. With the support of *CoQ10*, *resveratrol*, and *PQQ*, he nudged his cells towards renewal (Lee and Wei, 2005). Regular *movement, such as brisk walks and gentle strength work, became a ritual* not of exertion, but of cellular rejuvenation. Each session was a reminder to his body that energy was no longer something to fear.

He felt the subtle shift when his muscles didn't burn as quickly, when his breath became easier, and when meals stopped making him crash. His cells were remembering how to generate power.

To assist the process, James targeted *cellular energy metabolism* directly. He added *L-carnitine*, aiding the shuttle of fatty acids into mitochondria. He restored *thiamine (vitamin B1)*, supporting *pyruvate dehydrogenase* – a critical enzyme in ATP production (Wiley and Gupta, 2023). These weren't energy 'boosters'. They were *metabolic restorers* – turning the broken processes back on.

But energy could not be built on fire alone. Inflammation still flickered beneath the surface, whispering cytokine patterns that dulled receptor function and blocked conversion of thyroid hormones. So, he turned again to the gentlest healers: *omega-3 fatty acids*, rich in EPA and DHA, and *curcumin*, the golden root of calm. Together, they began to reduce the background hum of immune activation, allowing thyroid signals to land without distortion (Aggarwal and Harikumar, 2009).

He addressed *Th1/Th2 imbalance*, not through immune suppression but through *immune education. Vitamin D* and *zinc* became part of his protocol – not just to prevent deficiency, but to *re-teach the immune system tolerance*.

Next came the issue of *conversion* – the body's ability to take T4 and activate it into T3. James optimised *selenium and iodine* levels, both essential cofactors in thyroid metabolism. But he didn't stop there. He remembered that a large percentage of thyroid conversion happened not in the thyroid, but in the *gut*. He turned his focus inward, healing his microbiome, repairing intestinal integrity, and restoring *deiodinase activity* through probiotic support and digestive rest (Fasano, 2012).

He also returned to the cornerstone of endocrine repair: *cortisol balance*. Years of overstimulation had flattened his circadian rhythm, dulling his cortisol awakening response. With *ashwagandha* and *rhodiola*, he gradually supported his HPA axis back towards resilience. He paired these herbs with *mindfulness*, *sunlight in the morning*, *yoga in the evening*, and *tech-free hours before bedtime by avoiding cell phones or TV screens*. These became his tasks – gentle signals and easy habits of safety (Pruessner et al., 1999).

Finally, he addressed the *structure* – the very membranes that wrapped his mitochondria. Years of lipid peroxidation had made these membranes rigid, impeding hormone signalling. He reintroduced *omega-3 fatty acids*, not just for their anti-inflammatory effects, but to *restore fluidity*, allowing receptors to function, ions to pass, and energy to flow again (Stillwell, 2016).

Alongside these steps, James screened for *autoimmunity*. He discovered elevated thyroid antibodies and began adjusting his diet – removing gluten, reducing immune triggers, and embracing an *anti-inflammatory, gut-healing approach*. With time, his antibodies lowered. His inflammation softened. His energy returned – not in a flash, but in a quiet, steady glow (Fasano, 2012).

This chapter of his journey was not about 'fixing' the thyroid or taking a single supplement. It was about *rebuilding an ecosystem* – one where thyroid hormones and mitochondria could communicate again. Where receptors were sensitive, not resistant. Where energy was generated efficiently, used wisely, and no longer feared.

James wasn't just recovering. He was *repatterning*.

And in doing so, he wasn't just restoring metabolism. He was reigniting life.

ENZYMATIC CLEARANCE RESTORING TISSUE FLOW

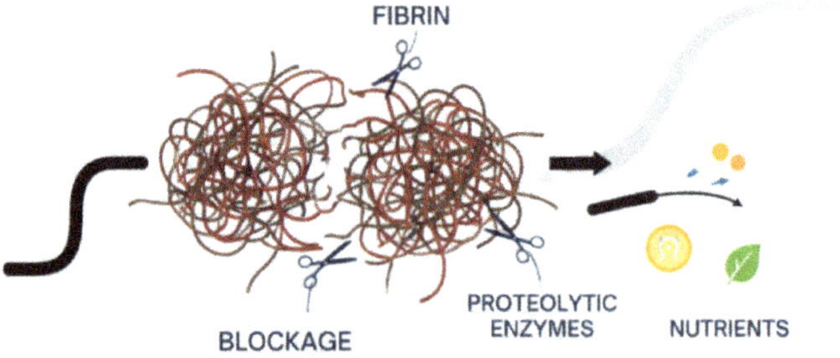

Proteolytic enzymes act like scissors, breaking down fibrin blocks to restore the flow of vital hormones and nutrients.

Chapter 16

Clearing the Way: Proteolytic Enzymes and Mitochondrial-Thyroid Receptivity

As James moved deeper into his healing, the insights from earlier chapters had begun to settle – not just as ideas, but as actions. His hormones were stabilising, his energy was returning, and his mitochondria were no longer so starved for signal. But still, something lingered: a thickness in the system; a subtle resistance. It was as if the path was open but *clogged with residue.*

That residue had a name: *fibrin*.

It had taken time to understand that his problem wasn't only hormonal or immune – it was also structural. As I described earlier, fibrin – a clotting protein designed to save lives in injury – had become a silent saboteur. In the face of chronic inflammation, his body had deposited it in places where it didn't belong – *around his capillaries, within connective tissues,* and *across the extracellular matrix*. These deposits acted like cobwebs in the bloodstream, *choking circulation*, *stiffening the cellular environment*, and *blocking thyroid hormones from reaching their targets*.

James had already begun repairing the biochemical landscape – antioxidants, mitochondrial nutrients, and T3 support. Now, he needed to *clear the physical barriers* to progress. That's when he discovered the potential of *proteolytic enzymes*.

Names like *Serrapeptase*, *Nattokinase*, *Lumbrokinase*, and *Bromelain* entered his vocabulary – not as supplements, but as *biological janitors*, capable of degrading the fibrin that had hardened his tissues and narrowed his capillaries. These enzymes were not stimulating or suppressing – they were *clearing*. Cleaning up what the body had left behind in its long war with chronic inflammation (ScienceDirect, 2005).

He began slowly. He started taking *Serrapeptase* in the morning, between meals. Then came *Nattokinase*, chosen for its vascular specificity. As he added these enzymes, he noticed subtle changes – *less brain fog*, *warmer extremities*, and *clearer vision in the afternoons.* His tissues were finally receiving the oxygen and nutrients they had been waiting for.

At the cellular level, he was undoing years of damage. *Fibrin clots*, small but persistent, had been reducing *microvascular flow*, limiting not only oxygen but also the delivery of *T3 to his peripheral tissues*. With fibrin broken down, *thyroid hormones could finally move freely*, crossing capillaries, reaching cells, and activating the *mitochondrial receptors* that governed energy production and metabolic rhythm (Wrutniak-Cabello et al., 2001).

There was more. The stiffened *extracellular matrix*, once bound up with fibrin, had been physically impeding the access of T3 to *membrane-bound thyroid receptors*. With the enzymes reducing matrix rigidity, *receptor binding improved*. Mitochondria, no longer compressed by oxidative debris and mechanical resistance, began to *breathe again*, responding to T3 with increased ATP production, fat burning, and thermogenesis. Energy, once a scarce resource, became available again – on demand.

He came to appreciate that these enzymes were not only improving circulation – they were also *calming inflammation*. Fibrin deposits often served as *platforms for immune cell activation*, creating chronic irritation in tissues. By clearing these deposits, James was not only restoring blood flow – he was also lowering *cytokine noise*, reducing the Th2-skewed signals that had plagued his body for years.

His recovery now had *structure as well as function.*

The benefits became cumulative:

- ***Improved circulatory health***, with less tissue hypoxia and more efficient hormone delivery.
- ***Enhanced mitochondrial activation***, as T3 finally arrived where it was needed.
- ***Reduction in systemic inflammation***, breaking feedback loops between fibrin, immune overactivation, and energy loss.

For James, this wasn't just another intervention – it was *clearing the final block.* The tunnel was built. The train was running. But the tracks had been clogged with debris. Now, with fibrin dissolving and signals moving, his systems could finally coordinate.

He began to feel a kind of lightness – not just in energy, but in movement, in breath, and in clarity. It wasn't a spike of wellness – it was *a steady return to flow.*

This wasn't quite the end of his journey – but it was the beginning of *unobstructed progress.*

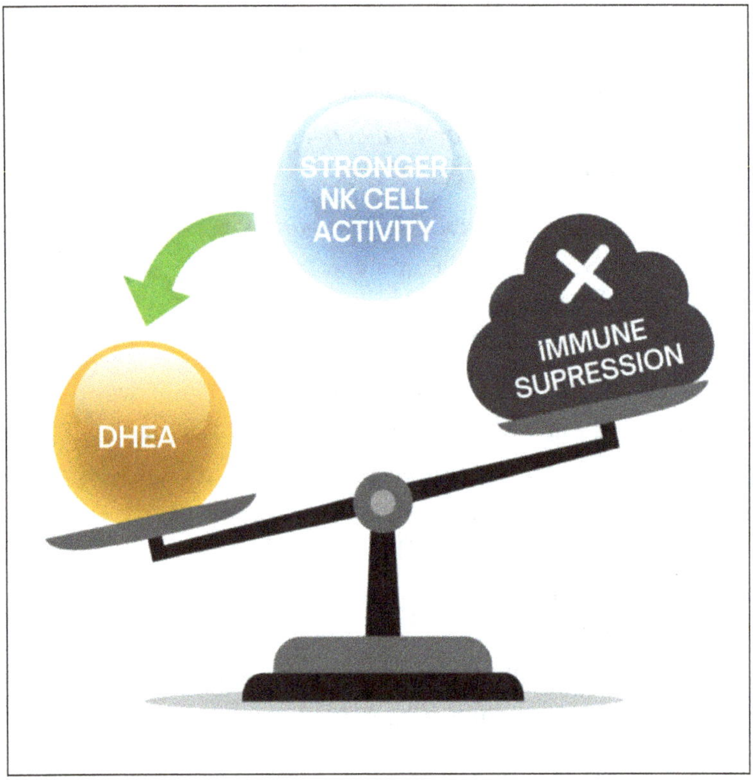
STRONGER
NK CELL
ACTIVITY
DHEA
IMMUNE
SUPRESSION

Chapter 17

Rebalancing the Scales: DHEA and the Rise of NK Cytotoxicity

By now, James had peeled back layer after layer of his condition – unravelling the hormonal, mitochondrial, and immunological threads that had kept his body in a prolonged state of defence. But there was still one piece of the puzzle that felt inert. One arm of his immune system that remained quiet for too long: *his natural killer cells.*

He had come to understand that NK cells weren't just sentinels for viral threats or cancer – they were *crucial regulators of immune tone*. In the face of *Th2 dominance*, their silence had allowed chronic infections to linger, inflammation to persist, and neutrophils to over-react. But now, James was exploring how to *wake them up – gently, strategically, and without overstimulation.*

That's when he found *DHEA*.

Once considered merely a 'youth hormone', *Dehydroepiandrosterone* was emerging in James' research as a *powerful immune modulator* – a hormone secreted by the adrenal glands with profound influence over *cytokine balance, NK cell function, and inflammatory signalling*. It was not a stimulant, but *a messenger of recalibration*.

He learned that DHEA had the ability to *enhance NK cell cytotoxicity*, primarily through the induction of *interleukin-12 (IL-12)*. This cytokine served as the ignition switch for NK activity – prompting the production of *IFN-γ,* a crucial signal that not only strengthened NK killing potential but also helped shift the immune profile *away from Th2 dominance and towards Th1 resilience* (Solerte et al., 1999; Sompayrac, 2022).

For James, this was the turning point. By supporting IL-12 and IFN-γ, DHEA didn't just reawaken his NK cells – it also *calmed the Th2 cascade*, reducing the overproduction of IL-4 and IL13, the very cytokines that had driven his allergies, his eczema, and his hypersensitive gut.

But the influence of DHEA didn't stop there.

It had a powerful regulatory effect on *IL-6*, one of the core drivers of chronic inflammation and neutrophil activation. In his own lab results, James had seen elevated IL-6 for years – linked to flares in fatigue, joint pain, and brain fog. What he had not known was that *DHEA downregulated IL-6 by suppressing NF-κB,* the transcription factor that commanded many of the body's inflammatory loops (Straub et al., 2010).

By doing so, DHEA interrupted a *critical axis of chronic inflammation*: the loop between IL-6, neutrophil activation, and tissue damage. The effect wasn't immediate, but over time James felt the difference. Less burning behind his eyes. Less puffiness in his joints after physical activity. Fewer episodes of mental drain after minor stressors. He was, finally, *interrupting the cycle*.

The deeper benefit was in the *balance it restored.*

DHEA didn't just activate Th1 responses or inhibit Th2. It *rebuilt the seesaw*. His immune system was no longer responding through a single channel – it was becoming multidimensional again. It could clear pathogens more efficiently without overreacting to harmless triggers. His *NK cells were now active*, not dormant. His *neutrophils were less over-recruited.* His *inflammation became more modulated.*

For the first time in years, James felt like his immune system was *self-correcting.*

He approached DHEA supplementation cautiously – checking

levels, titrating slowly, and working with a practitioner who understood the nuance. With each step, he felt more stable, energised, but also *resilient*.

This was not the superficial kind of energy that came from caffeine or adrenaline. It was deeper. Quieter. It felt as if his body had remembered *how to hold its ground*, rather than constantly reacting or collapsing.

The restoration of NK cytotoxicity was not about aggression – it was about *clarity*. Clarity of signal. Clarity of self-recognition. Clarity of what should be eliminated and what should be left alone.

This clarity, James thought, *was the beginning of true immunity*.

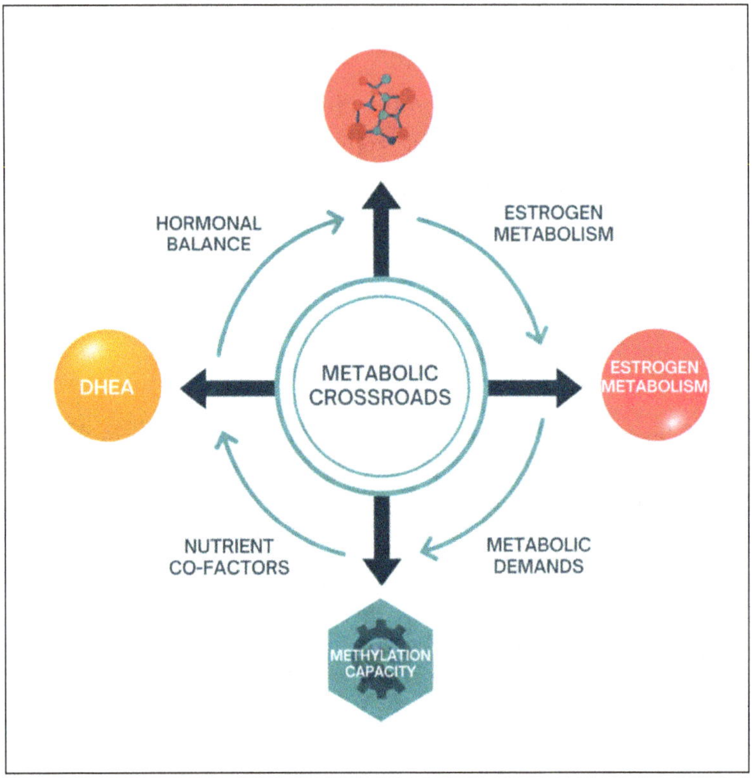
HORMONAL
BALANCE
ESTROGEN
METABOLISM
DHEA
METABOLIC
CROSSROADS
ESTROGEN
METABOLISM
NUTRIENT
CO-FACTORS
METABOLIC
DEMANDS
METHYLATION
CAPACITY

Chapter 18

Hormonal Crossroads: DHEA, Methylation, and the Return of Oestrogen's Shadow

James had long known the face of *oestrogen dominance*. He had felt it in his skin – when inflammation flared; in his gut – when histamine reactions spiked; and in his energy – when thyroid signals hit walls of resistance. He had already explored the role of impaired detoxification, sluggish methylation, and the oestrogen–thyroid–mitochondria triad. But he hadn't expected *that DHEA, something which was designed to support his healing – might also risk pulling him back towards that hormonal imbalance.*

Yet, as he continued with his carefully titrated DHEA regimen – one that had revived his NK cells, restored part of his Th1 function, and calmed the hyperactive Th2 profile – he began to feel the *subtle return of a familiar pattern.*

His energy, which had begun to stabilise, took a strange turn – which can be described as *tired but wired.* His sleep pattern changed, becoming lighter; his skin felt reactive again; and his core temperature dipped, despite thyroid support. And most telling of all, his *free T3 levels began to slide*, even though his medication and mitochondrial cofactors hadn't changed.

He knew the signs. *Oestrogen was creeping upward again.*

DHEA, while immune-supportive, *fed into the same pathways* that had once overburdened his hormonal metabolism. As an adrenal precursor, it could be converted to androgens, and – via *aromatase* –

into *oestradiol*, the most biologically active oestrogen. In a body like James', with a history of methylation dysfunction and inflammatory triggers, *aromatase expression was already heightened*, especially in adipose tissue and under chronic immune load (Zhu et al., 2021).

What followed was predictable, but still frustrating. *Oestrogen elevation increased thyroid-binding globulin (TBG)*, reducing the amount of free T3 available for mitochondrial receptors. The signals were still present – but muffled, trapped in transit. And worse still, rising oestrogen fed back into oxidative stress, *damaging mitochondrial membranes*, and further *dampening T3 receptor sensitivity* (Cloyd, 2025).

James had already worked hard to *repair these pathways once;* he wasn't about to let them slip again.

So rather than abandoning DHEA, he chose to adapt. *Balance, not elimination*, would be the path forward.

He began with the *methylation pathways*, knowing that they were the backbone of oestrogen clearance. He rechecked his nutrient baselines and increased his intake of *methylated folate, B12, and betaine (TMG)* – not just for neurotransmitter support, but to aid in *COMT-mediated oestrogen deactivation*. These were not optional – they were required for *moving oestrogen out of the system efficiently*, reducing its burden on the thyroid and mitochondria.

Next, he turned again to the liver. He reinforced his Phase I and Phase II detoxification using *DIM* from cruciferous vegetables, *Calcium D-glucarate* to block beta-glucuronidase activity, and a daily intake of *sulphur-rich foods* like onions, garlic, and arugula to support *glutathione conjugation* and safe oestrogen excretion.

But he didn't stop at detox. He knew now that to *regulate aromatase*, he had to *modulate the terrain*. That meant addressing the *copper–zinc balance*, since high copper and low zinc favoured oestrogen synthesis. It also meant introducing *flavonoids* – such as *apigenin and quercetin* – which gently inhibited aromatase activity without pushing his system too far in the opposite direction.

This was a nuanced process – *watching ratios, not just levels.*

And it was working. His oestrogen symptoms began to fade, not abruptly, but gradually. His temperature normalised. His energy grew steadier. And most importantly, his labs showed *a more favourable T3 to rT3 ratio* – a sign that the thyroid hormone was finally reaching its destination again.

James didn't demonise DHEA for this hormonal detour. Instead, he saw it for what it was: *a powerful tool*, but one that needed *context*. In a system like his – one with a past marked by oestrogen overload and sluggish detox – every intervention had to be supported by infrastructure. DHEA could help rebuild immunity and enhance mitochondrial signalling, *if paired with the right metabolic scaffolding.*

This chapter in his recovery was not about adding more – it was about *refining what was already working*, tightening the feedback loops, and protecting progress from regression.

What DHEA had taught him – more than anything – was that *every hormone is a message*, and every message must be read in the context of the whole.

Healing was not about pushing any one system harder. It was about *knowing when to slow down, when to support, and when to step back and recalibrate.*

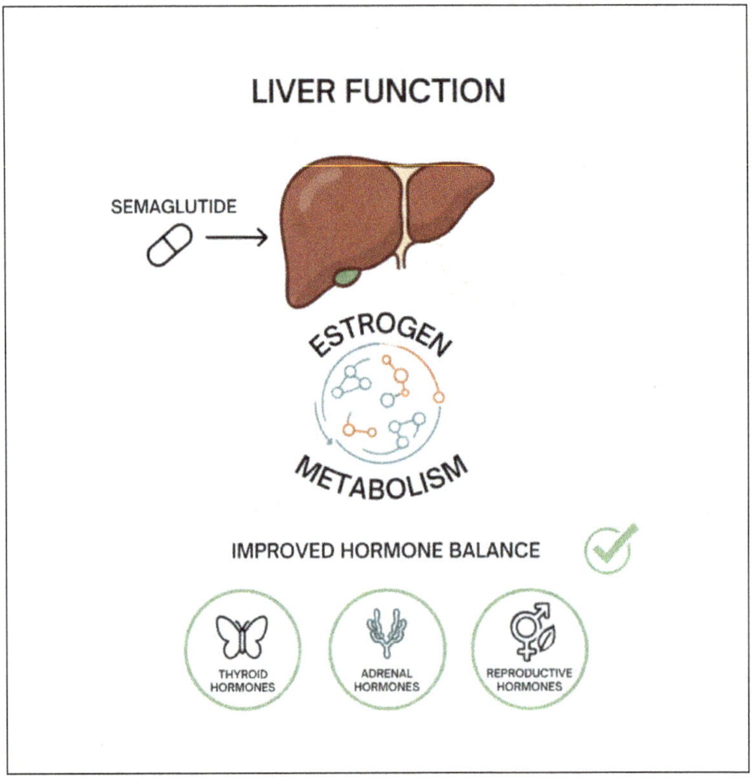
LIVER FUNCTION
SEMAGLUTIDE
ESTROGEN
METABOLISM
IMPROVED HORMONE BALANCE
THYROID HORMONES
ADRENAL HORMONES
REPRODUCTIVE HORMONES

Chapter 19

A New Path Through the Liver: Semaglutide, TBG, and the Resolution of Oestrogen-Driven Immune Skew

James had learned, sometimes the hard way, that endocrine recovery was a game of nuance – not dominance. What seemed to help one system could tip another. DHEA, so helpful for his immune system, had nearly reignited his oestrogen burden. Methylation pathways, so vital for detoxification, needed constant fuelling. Every shift brought clarity, but also new complexity.

And now, the spotlight turned again – this time, back to the liver.

He had long suspected that *oestrogen dominance* was not just a hormonal imbalance, but a *metabolic terrain*. Elevated oestrogens had increased his *thyroid-binding globulin*, leaving his *free T3* levels suppressed. His lab work continued to reflect a frustrating contradiction: *normal TSH, normal total T3 – but low free T3 and high reverse T3*. The signals were present, but unavailable. His mitochondria, ready for instruction, weren't receiving the call.

This wasn't a thyroid issue alone. It was *a liver issue*.

Oestrogens signalled the liver to produce more TBG. And TBG, in turn, tightly bound his circulating thyroid hormone, making it *unavailable to cells*. The consequence was *metabolic sluggishness*, but also a more profound immune effect – *a weak-*

ening of mitochondrial support that left his immune system prone to Th2 dominance once again.

But this time, James came across an unexpected ally: *semaglutide*.

Initially designed as a medication for type 2 diabetes and weight management, semaglutide worked by activating *Glucagon-Like Peptide-1 (GLP-1) receptors* – a mechanism that had profound effects on insulin, appetite, and glucose control. But as James soon discovered, the effects went *far beyond blood sugar*. Semaglutide improved *hepatic insulin sensitivity*. It reduced *hepatic steatosis* – the fatty infiltration of the liver that interfered with detox and hormone regulation. And critically, it had been shown to *lower hepatic TBG production* by addressing both *oestrogenic overload and insulin resistance* (Drucker, 2018). This was what James needed – not a hormonal blocker, but a *metabolic corrector*.

He began, under clinical supervision, titrating his dose carefully. Within weeks, he noticed the shift – not in weight, but in *clarity*. His energy was more stable and his post-meal crashes less dramatic. And as the fog began to lift, he saw the deeper benefit: his *free T3 levels began to rise*.

By reducing TBG production, semaglutide was *freeing thyroid hormone from its biochemical shackles*, allowing it to *reach receptors*, activate mitochondria, and recalibrate energy metabolism. The cascade was immediate and cellular: *T3 responsiveness improved*, together with mitochondrial tone and thermogenesis, leaving a sense of being *re-engaged with life's rhythm*.

But semaglutide was not working through the thyroid alone. It was also *modulating his immune system*.

James learned that *GLP-1 receptors exist on immune cells*, not just in the pancreas. Their activation had been shown to *suppress pro-inflammatory cytokines* – including *IL-6 and TNF-α*, the very molecules that had driven his Th2 dominance and mitochondrial dysfunction for years (Drucker, 2018). As inflammation lowered, so too did the *Th2 skew* that had defined his immune profile.

This was not immune suppression – it was *rebalancing*. A quieting of the overreaction, a lifting of the inflammatory veil.

What emerged, for the first time in years, was *clarity at every level*:

- ***Metabolic clarity***, with restored insulin signalling and energy production.
- ***Hormonal clarity***, with reduced TBG and liberated T3.
- ***Immunological clarity***, with lowered cytokine noise and a return to Th1/Th2 equilibrium.

And perhaps most importantly, *neurological clarity*. The brain fog that once ruled his afternoons was fading. His capacity to plan, to feel, to rest – was returning.

James didn't see semaglutide as a miracle pill. He saw it as a *strategic metabolic modulator*, working through the pathways that had quietly distorted his hormonal and immune systems. Its strength lay not in direct suppression, but in *unblocking the metabolic bottlenecks* that made hormone signalling and immune coherence impossible.

In that way, semaglutide was not correcting a symptom; it was *releasing the system's own intelligence*.

And in doing so, it brought James one step closer to what he had been seeking all along – not just the absence of disease, but the presence of *physiological trust*.

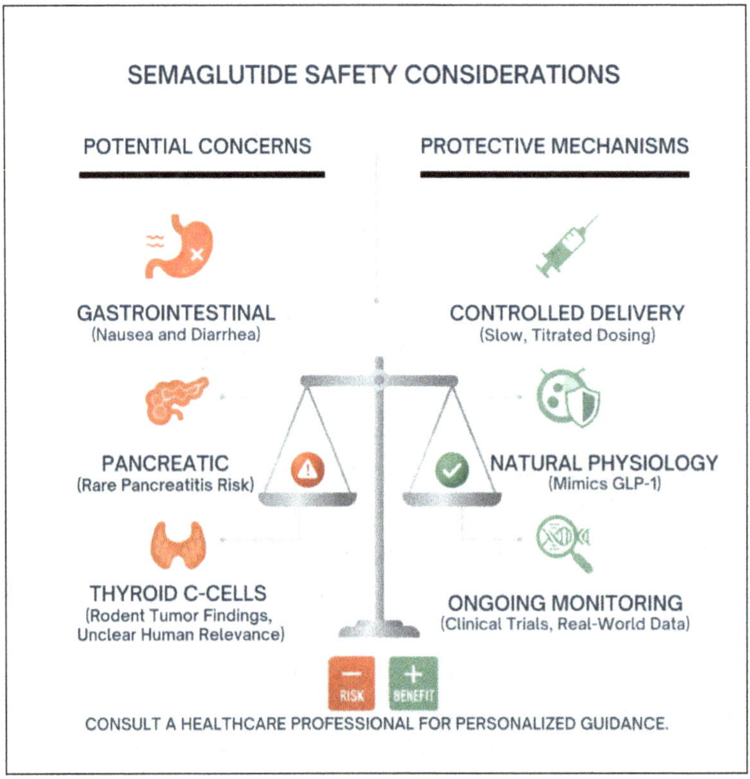
SEMAGLUTIDE SAFETY CONSIDERATIONS
POTENTIAL CONCERNS
PROTECTIVE MECHANISMS
GASTROINTESTINAL
(Nausea and Diarrhea)
CONTROLLED DELIVERY
(Slow, Titrated Dosing)
PANCREATIC
(Rare Pancreatitis Risk)
NATURAL PHYSIOLOGY
(Mimics GLP-1)
THYROID C-CELLS
(Rodent Tumor Findings,
Unclear Human Relevance)
ONGOING MONITORING
(Clinical Trials, Real-World Data)
RISK
BENEFIT
CONSULT A HEALTHCARE PROFESSIONAL FOR PERSONALIZED GUIDANCE.

Chapter 20

Navigating the Unknown: Semaglutide and the Cancer Question

James had learned to be cautious.

Each intervention he embraced was preceded by research, reflection, and instinct. Healing was not about chasing solutions; it was about making peace with uncertainty while carving a path through it. And as semaglutide became a cornerstone in his metabolic recovery, a new question surfaced – *what about pancreatic cancer?*

He didn't ask this question out of paranoia. His history had taught him that *every gain had to be earned, understood, and integrated.* The progress he had made with semaglutide was undeniable – his energy was more stable, his insulin sensitivity had improved, and his free T3 had finally begun to rise. But lingering in the background was a narrative – news stories, scattered forums, and warnings on medication inserts – all hinting at a possible link between GLP-1 medications and pancreatic risk.

James paused – not out of fear, but to investigate.

The origin of the concern, he discovered, stemmed from *animal studies.* In certain rodent models, high doses of GLP-1 receptor agonists had triggered *pancreatic ductal hyperplasia*, increased pancreatic mass, and signs of *low-grade inflammation* – changes that some researchers feared could precede malignancy. The mechanism appeared to involve prolonged activation of GLP-1 receptors, raising intracellular cAMP (cell signalling) and altering pancreatic signalling pathways.

But the more James learned, the more critical *context* became.

Rodents *have a much higher density of GLP-1 receptors in pancreatic tissues* than humans. Their response to these medications – particularly at pharmacological doses – was *exaggerated by species-specific receptor biology*. What had been observed in mice simply had not translated to other species – not in *non-human primates*, and crucially, *not in humans* (Drucker, 2018).

Human studies told a different story.

Large-scale cohort analyses, including one recent study tracking over half a million patients, *showed no increased risk of pancreatic cancer* among GLP-1 users. Meta-analyses of randomised controlled trials – the gold standard of clinical research – also found *no significant association*. The regulatory body *US Food and Drug Administration* (*FDA*) had conducted extensive post-marketing surveillance, and while they committed to continued monitoring, they found *no causative link* between GLP-1 RAs and pancreatic cancer (FDA, 2020).

The discrepancy, James realised, was not uncommon. Science was full of cautionary animal studies – some relevant, some not. The key was *contextual interpretation*, not reaction.

What mattered most to him was the mechanism.

The concern in rodents had been tied to *exocrine pancreatic stress* – changes in the *ductal system*, not the insulin-producing islets. But semaglutide acted primarily on *GLP-1 receptors in the brain, gut, and beta cells* – modulating appetite, insulin release, and glucose control. The *expression of these receptors in human exocrine tissue was minimal*, and no increase in pancreatitis or neoplasia had been consistently reported in long-term clinical data.

Still, James chose to be cautious. Not fearful, but proactive.

He worked with his clinician to track *lipase and amylase levels* regularly – not because they were predictive of cancer, but because *pancreatic inflammation* could be an early sign of imbalance. He continued to monitor for *digestive changes*, *abdominal discomfort*, or unexplained shifts in energy or blood sugar. And most importantly, he reminded himself of the broader terrain.

Cancer was not about one receptor or one drug. It was about *terrain* – about chronic inflammation, oxidative stress, metabolic dysfunction, and immune surveillance. These were the very things *semaglutide was helping him improve*. By lowering insulin resistance, reducing hepatic fat, modulating IL-6, and restoring mitochondrial tone, he was *actively lowering his cancer risk profile*, not increasing it.

James also knew that *health was not a guarantee – it was a negotiation*. A relationship between risk and benefit, between mechanism and meaning. Semaglutide had given him back his metabolic rhythm, had unburdened his liver, and had unblocked thyroid signals that had been trapped for years. It had brought him into the *kind of clarity and stability he hadn't known since his twenties*.

That mattered.

And as he walked forward, he did so not with blind faith, but with *informed vigilance*. He would continue to watch the literature. He would continue to listen to his body. But he would not let *unfounded fear override the progress he had earned.*

For James, healing was never about eliminating risk.

It was about *making peace with it – wisely, attentively, and fully awake*.

24-HOUR CIRCADIAN RHYTHM:
NIGHTLY REST & IMMUNE HEALTH

Melatonin promotes restorative sleep, crucial for immune defense.

Chapter 21

The Rhythm Restored: Melatonin and the Softening of Th2 Dominance

By now, James had come to understand that healing was not only about action – it was also about *timing*.

He had the sense that there still existed a rhythm – *a biological cadence that had long been misaligned*, one that regulated energy, inflammation, sleep, and immunity in tandem.

At the heart of this rhythm was *melatonin* – a molecule he had once thought of as merely a sleep aid, but now understood it as a *multisystemic modulator*: a synchroniser of clocks, a protector of cells, and a *quiet conductor* of immune balance.

He began to see that melatonin played a critical role in correcting the *Th2 dominance* that had defined much of his chronic immune dysregulation. Research showed that it could directly *suppress IL-4 and IL-13*, the core cytokines responsible for the allergic, hyper-responsive state he had battled for years. It worked by *modulating NF-κB*, the transcription factor that governed much of his body's inflammatory reactivity (Carrillo-Vico et al., 2005).

This wasn't just chemical suppression – it was a signal to recalibrate.

Melatonin also functioned at a mitochondrial level. As an antioxidant, it moved effortlessly across cell membranes, *scavenging reactive oxygen species*, reducing lipid peroxidation, and even *enhancing antioxidant enzyme systems* such as superoxide dismutase and glutathione peroxidase (Lieurance, 2022). In mitochondria that were weighed down by oxidative stress and immune confusion, melatonin brought breath, balance, and restoration.

What struck James most was melatonin's influence on *T regulatory cells* – those silent architects of immune tolerance. In Th2-dominant states, Treg activity was often suppressed, leading to increased autoimmunity and chronic inflammation. Melatonin, however, promoted *Treg differentiation and function*, helping restore the checks and balances that had long been missing (Won et al., 2021).

Even more profound was melatonin's link to *circadian rhythm*. The immune system was deeply tied to the body's 24-hour clock. Immune cell migration, cytokine expression, and mitochondrial activity all followed *diurnal patterns* – patterns that were easily disrupted by stress, blue light exposure, irregular meals, or emotional dysregulation. Circadian misalignment had worsened his Th2 skew, while perpetuating mitochondrial dysfunction and hormonal disarray.

Melatonin didn't just put him to sleep – it *reset the rhythm*.

Paired with *morning sunlight*, reduced evening screen time, and deepened breath practices, it re-established a healthy divide between cortisol and melatonin, action and rest, alertness and healing (Cajochen et al., 2003). As his circadian rhythm rebalanced, he noticed he had deeper sleep, smoother mornings, and a nervous system that no longer braced against the day.

There was also its effect on *pro-inflammatory cytokines* like IL-6 and TNF-α – molecules that had long kept his immune system in a loop of low-grade, unresolving inflammation. Melatonin reduced their expression, breaking cycles that previously no supplement or diet could touch (Lieurance, 2022).

This wasn't another intervention to manage symptoms. It was *a regulator of the terrain itself* – a molecule that spoke the language of time, energy, and tolerance.

As James moved through this final arc of immune recalibration, he no longer chased 'fixes'. He was now learning how to *listen to rhythm*, to *align with physiology*, and to build a life that cooperated with the quiet intelligence of his cells.

Melatonin was no longer just a molecule of sleep. It was a *hormone of coherence*, of integration, of return.

In the deepening dark of each night, James no longer feared stillness.

He welcomed it.

Because now, his body *trusted it.*

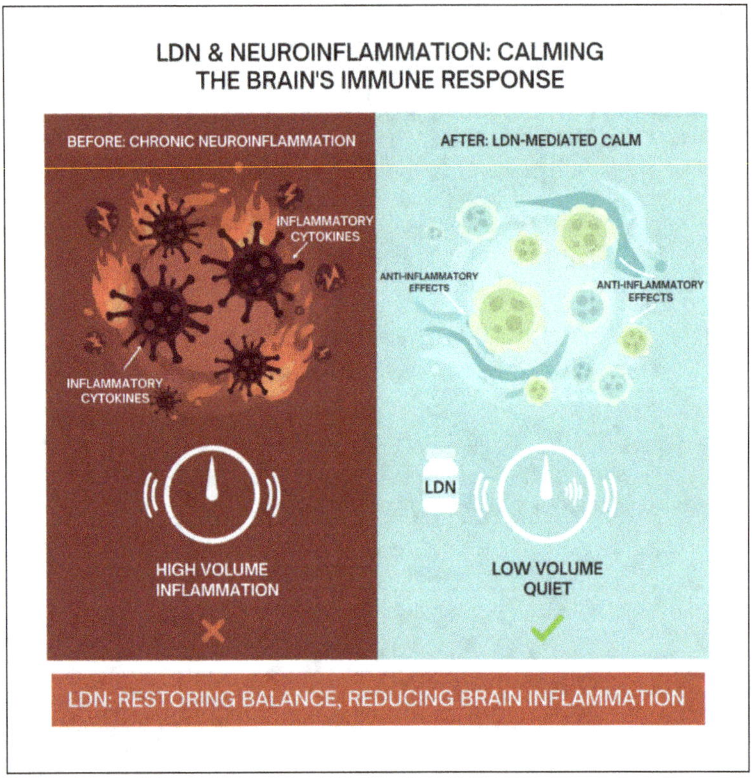
LDN & NEUROINFLAMMATION: CALMING THE BRAIN'S IMMUNE RESPONSE
BEFORE: CHRONIC NEUROINFLAMMATION
AFTER: LDN-MEDIATED CALM
INFLAMMATORY CYTOKINES
INFLAMMATORY CYTOKINES
ANTI-INFLAMMATORY EFFECTS
ANTI-INFLAMMATORY EFFECTS
LDN
HIGH VOLUME INFLAMMATION
LOW VOLUME QUIET
LDN: RESTORING BALANCE, REDUCING BRAIN INFLAMMATION

Chapter 22

Quiet Recalibration: Low-Dose Naltrexone and Immune Balance

James had come to see his immune system as a conversation – one that had grown louder and more chaotic over time. At first, the volume was manageable: a rash here, a reaction there. But eventually, the signals became distorted. What should have been regulatory became reactive. What should have been protective turned inflammatory. His immune system was not aggressive. It was *overwhelmed and misdirected.*

Even with the major systems now better aligned, James still sensed *a residual imbalance*. A subtle but persistent *inflammatory echo existed*, especially when stress crept in or sleep waned. It wasn't loud, but it was there. A low hum of immune agitation that had not fully resolved.

That's when he discovered *low-dose naltrexone (LDN)*.

At first, it seemed counterintuitive. Naltrexone? The opioid blocker used in addiction therapy? But LDN – *in a fraction of the dose traditionally used* – had an entirely different mechanism. It didn't suppress. It *rebalanced.* It worked not through brute force, but through gentle *immune modulation.*

He learned that LDN temporarily *blocked opioid receptors* – but only for a short time. This transient blockade triggered the body to *increase production of its own endogenous opioids*: *endorphins, enkephalins, and dynorphins* – molecules not only linked to mood and pain relief, but also to *immune regulation*, particularly through *Treg enhancement and NK cell support.*

This was the first layer of its magic: *a rise in natural opioids, leading to improved immune tolerance and surveillance*. James felt the shift not in stimulation, but in *stability* – less variability in his symptoms, and more coherence in his baseline.

The second mechanism was more targeted – and more profound. LDN modulated a lesser-known but powerful system: the *Opioid Growth Factor (OGF)–OGF receptor (OGFr) axis*. This pathway, driven by *[Met5]-enkephalin*, helped regulate *cell proliferation and immune cell activity*. By nudging this axis, LDN slowed the *excessive cell turnover* seen in autoimmune flare-ups and promoted *immune precision* rather than overreaction.

It was as if LDN didn't just calm the fire – it taught the immune system *how not to light it in the first place*.

But for James, the most transformative effect came through LDN's impact on the *brain's immune cells*: *the microglia*. In Th2 dominance and CDR1 states, these cells often became *chronically activated*, releasing cytokines like *IL-6*, *IL-12*, and *TNF-α*, creating a loop of inflammation, neuroinflammation, and downstream mitochondrial damage. LDN interrupted this loop by *antagonising Toll-like receptor 4 (TLR4)* – effectively telling the microglia to *stand down*.

And they listened.

Over weeks, James noticed a quieting – not just in his body, but also in his mind. The reactivity he had once felt after a poor night's sleep, or emotional stress, was now softer. His baseline mood was steadier, and his focus clearer. This was not a mood drug. Through *neuroimmune modulation*, it was *restoring his neurological environment*.

LDN was also quietly nudging his immune profile. His labs began to reflect a *reduction in IL-6*. And more importantly, he no longer felt dominated by *Th2-patterned responses*. His sinuses, his skin, his gut – systems that had once acted as alarm bells – were *less reactive*. His immune system had become more *intelligent, more discerning*.

LDN wasn't replacing anything – it was *amplifying the body's own regulatory networks*.

For James, this meant three profound shifts:

1. ***Immune tolerance was returning***, allowing him to interact with the world without constant reaction.

2. ***Neuroinflammation was subsiding***, creating space for mood and cognition to stabilise.

3. ***Th1/Th2 balance was being restored***, not by suppressing Th2, but by supporting *true immune orchestration*.

He saw why LDN had been used successfully in a range of conditions – *autoimmunity, allergies, chronic pain, and even cancer*. Not because it cured them, but because it addressed something upstream: *the tone of the immune system*, the tempo of its response.

And in that sense, LDN was not an intervention. It was a *recalibration*.

The whisper of inflammation that had once filled every quiet moment was fading.

What remained was space.

Space for healing. Space for presence. Space to rebuild not just systems – but *trust in the body's wisdom*.

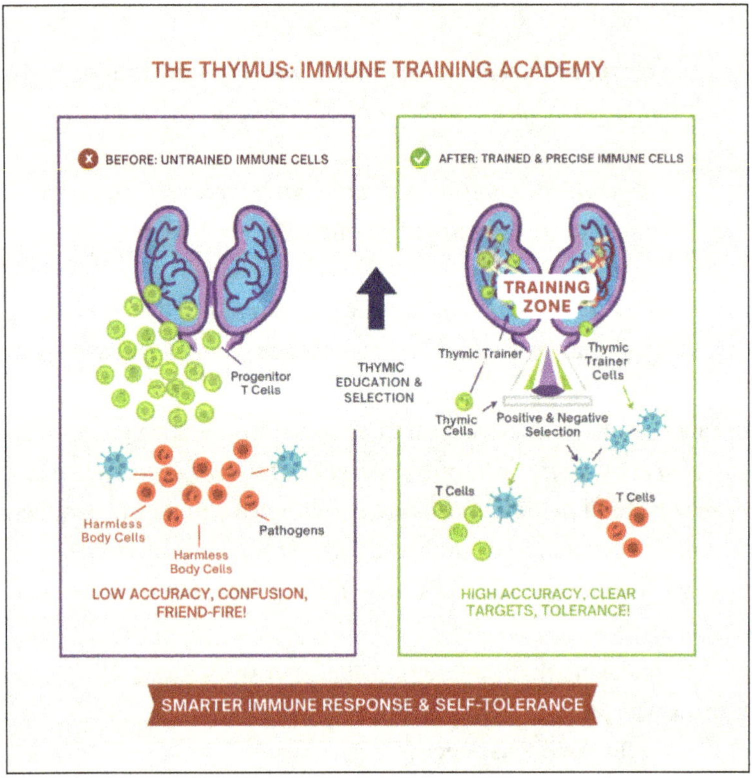
THE THYMUS: IMMUNE TRAINING ACADEMY
BEFORE: UNTRAINED IMMUNE CELLS
Progenitor T Cells
Harmless Body Cells
Harmless Body Cells
Pathogens
LOW ACCURACY, CONFUSION, FRIEND-FIRE!
THYMIC EDUCATION & SELECTION
AFTER: TRAINED & PRECISE IMMUNE CELLS
TRAINING ZONE
Thymic Trainer
Thymic Trainer Cells
Thymic Cells
Positive & Negative Selection
T Cells
T Cells
HIGH ACCURACY, CLEAR TARGETS, TOLERANCE!
SMARTER IMMUNE RESPONSE & SELF-TOLERANCE

Chapter 23

Rewriting the Immune Narrative: Thymosin Alpha-1 and the Return of Cellular Precision

By now, James had come to understand that his immune system was not broken, but rather *miscalibrated*. A system that once protected him had become confused – skewed towards the reactive, driven by chronic Th2 signalling, and starved of the clear, cell-mediated responses that once kept viruses, inflammation, and cancer in check.

He had corrected much of the terrain and rebuilt his immune tolerance through LDN, melatonin, and metabolic repair. But something still felt unfinished.

The deeper layers of immunity – the *innate bridges to adaptive clarity* – had not yet fully returned.

That's when James was introduced to *thymosin alpha-1 (Tα1)*.

It was not a hormone in the conventional sense. It was a *peptide* – a small, naturally occurring protein fragment produced by the thymus gland, a forgotten but powerful organ in the architecture of immune education. With age, stress, and chronic inflammation, the thymus involutes. So too does the clarity of immune messaging.

Tα1, it turned out, could help restore that lost clarity.

Its first action was to *promote Th1 differentiation*. In James' immune system – long skewed towards Th2 cytokines like IL-4 and IL-13 – this meant a return of *IFN-γ* and *IL-2*, the foundational Th1 messengers that support *intracellular pathogen clearance*, *tumour surveillance*, and the *immune tone of resilience*.

For years, James had lacked this signature. His body responded to the world through skin reactions, sinus congestion, food intolerance – *all outward signs of a system defaulting to Th2*. But with Tα1, the inward-directed, *cell-mediated branch of immunity* – the one that quietly cleared viruses, repaired damaged tissue, and maintained immune memory – began to return.

Tα1 also worked through another forgotten frontier: the *dendritic cell*. These antigen-presenting cells serve as translators between the innate and adaptive systems. In a Th2-dominant state, dendritic cells often remain *immature or underactivated*, leading to fuzzy signalling and a preference for tolerance over surveillance. But with Tα1, dendritic cells matured, became more responsive, and *improved the handoff to T cells*, especially in favour of Th1 outcomes.

The immune conversation was being restored.

Another critical piece was the *enhancement of natural killer cell activity* – something James had worked towards through DHEA, melatonin, and other modulators. But Tα1 brought new precision. It activated NK cells not through brute stimulation but by *rebalancing intracellular signalling*, allowing these first-responder cells to regain their cytotoxic clarity.

The result was not hyperimmunity. It was *targeted immune surveillance* – a return to *silent defence*, where viral remnants and misbehaving cells were cleared without triggering waves of collateral inflammation.

James noticed something else: *his tissues began to settle*. His skin felt less volatile. His gut less reactive. His thoughts sharper. This was not only immune enhancement – it was *immune restoration*.

At the molecular level, Tα1 modulated *Toll-like receptors (TLRs)* – those ancient pattern recognisers of the immune system. Through TLR engagement, Tα1 helped *activate NF-κB and IRF3*, signalling cascades that supported *early antiviral and antitumour responses* while reducing *chronic inflammatory noise*. This dual action – calming the wrong alarms while reigniting the right ones – was exactly what James' system needed.

And because of its broad reach, the therapeutic implications of Tα1 were equally vast:

- ***Autoimmunity***, where Th2 dominance creates chaotic antibody responses, saw improved regulation.
- ***Chronic viral infections***, once invisible behind suppressed NK activity, became more manageable.
- Even ***oncological applications*** were supported – through Tα1's role in improving tumour surveillance and potentiating immunotherapy responses.

But for James, it wasn't just about mechanisms. It was about *feeling immune confidence for the first time in years*. He was no longer reacting to pollen, food, or stress; no longer bracing for another wave of fatigue after a simple cold.

This was his immune system *remembering what it once was* – silent, strong, and sovereign.

Tα1 had not 'stimulated' anything. Rather, it had *reminded the system of its original instructions*.

And in that remembering, James found his own return – not to the body he once had, but to *a new relationship with it*. One grounded in communication, clarity, and cellular truth.

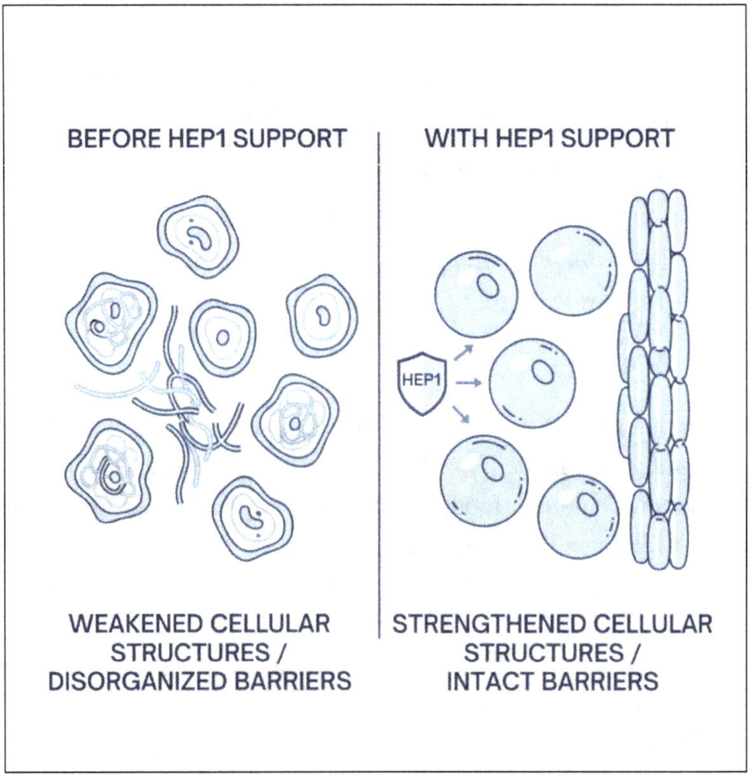
BEFORE HEP1 SUPPORT
WITH HEP1 SUPPORT
HEP1
WEAKENED CELLULAR STRUCTURES / DISORGANIZED BARRIERS
STRENGTHENED CELLULAR STRUCTURES / INTACT BARRIERS

Chapter 24

Structure Becomes Signal: Human Ezrin Peptide 1 and the Final Phase of Immune Restoration

James had come to learn that healing was not just about balancing signals – it was also about rebuilding *infrastructure*. For years, his immune system had been dysregulated not only because of biochemical imbalances, but because its *architecture had collapsed under the weight of chronic inflammation.*

The deeper he looked, the more he realised that *structure and function were one and the same*. It was not enough to modulate cytokines or recalibrate T cell responses. The *immune landscape itself – its scaffolding, its networks, its communication hubs – needed repair*.

That is when he encountered the quiet potential of *Human Ezrin Peptide 1 (HEP1)*.

At first glance, HEP1 – an emerging therapy – did not seem like a headline-grabbing molecule. But as James delved into the literature, he realised it wasn't a stimulant or suppressant – it was a *builder;* a peptide that *restored immune intelligence not through force, but by supporting the structural integrity of the immune system itself.*

One of the first things HEP1 did was *restore lymph node architecture*. For years, chronic Th2 dominance had created low-grade lymphatic inflammation, degrading the very networks responsible for effective antigen processing. HEP1, by *enhancing fibroblast activity and extracellular matrix remodelling*, helped *rebuild these immune hubs* – supporting antigen presentation, adaptive immune engagement, and the regeneration of immune memory.

It was the immune equivalent of *repairing the city's post office before expecting the mail to be delivered on time*.

James could feel the difference – not as an adrenaline rush, but as *improved response fidelity*. His immune system wasn't firing wildly anymore. It was beginning to act with *precision and context*.

HEP1 also worked at the *B cell level*, engaging with the *ezrin-actin cytoskeletal complex* – a structure involved in *receptor clustering and signal transduction*. By optimising B cell receptor (BCR) activity (B cells produce antibodies), HEP1 supported *more targeted antibody responses*, rather than the broad, excessive humoral activity that characterised his previous Th2-dominant state.

For James, this translated into a *reduction in hypersensitivities*. Foods that had once caused flushing or fatigue no longer triggered a storm. His sinus congestion – chronic for decades – began to subside. These were signs of an immune system that was no longer guessing. It was *remembering*.

HEP1's impact on inflammation was equally profound. Through *NF-κB inhibition*, it downregulated *IL-6* and other inflammatory cytokines, many of which had kept James locked in a feedback loop of low-grade systemic inflammation. With those signals quieted, *Treg activity increased*, and the *Th1/Th2 balance finally began to stabilise*.

James noticed this change mostly in his *nervous system*. Where he had once been easily overwhelmed, he now found he could *move through stress without crashing*. His body wasn't responding as if every trigger was a threat. It was *processing, resolving, and restoring* – hallmarks of an immune system that had returned to trust.

And then came the final layer: *immune amplification – on purpose*.

HEP1 functioned not only as a restorative peptide but as a *gentle immune adjuvant*. It enhanced the *specificity and magnitude of T cell responses*, making it ideal for both *vaccine support* and *immunotherapy adjunctive care*. For James, who had a long history of sluggish responses to vaccines and unresolved viral exposures, this was the missing link – *a way to wake up targeted immunity without risking systemic overdrive*.

HEP1 was even showing promise in *oncological research*, through its ability to *restore effective immune surveillance and reduce tumour immune evasion*. It was not an anti-cancer drug – it was a *signal amplifier*, helping the immune system return to its original design: to *detect, respond, and resolve*.

For James, HEP1 became the link between *immune recovery and long-term immune intelligence*.

It didn't just regulate. It *restored function through structure*.

By rebuilding lymphatic architecture, remodelling antigen presentation, and refining immune clarity, HEP1 marked the final leg of James' return – not just to health, but to *physiological coherence*.

His immune system was no longer at war with itself. It was a team again – rooted, resilient, and ready.

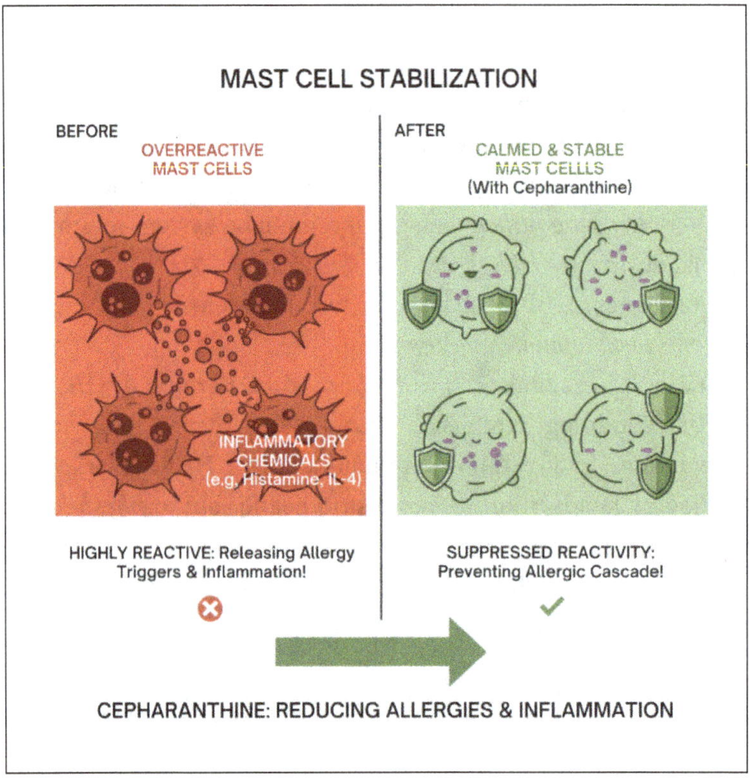
MAST CELL STABILIZATION
BEFORE
OVERREACTIVE MAST CELLS
AFTER
CALMED & STABLE MAST CELLLS
(With Cepharanthine)
INFLAMMATORY CHEMICALS
(e.g. Histamine, IL-4)
HIGHLY REACTIVE: Releasing Allergy Triggers & Inflammation!
SUPPRESSED REACTIVITY: Preventing Allergic Cascade!
CEPHARANTHINE: REDUCING ALLERGIES & INFLAMMATION

Chapter 25

The Hidden Architecture: Cepharanthine and the Cellular Resolution of Th2 Dominance

James had walked a long road. What had begun as a search for answers – why his immune system overreacted, why his energy collapsed, and why inflammation whispered through every tissue – had evolved into something more profound: *a reconstitution of inner order.*

He had quieted inflammatory cytokines, unblocked hormone pathways, recalibrated the mitochondria, and rebalanced Th1/Th2 dynamics through immune modulation. Yet beneath it all, there remained a kind of cellular fragility – a sense that *his immune system was still easily pushed off course*, still vulnerable to the micro-instabilities that once triggered full-body flares.

That's when he came across *Cepharanthine* – a compound he had never heard of before. A *plant-derived alkaloid*, long used in Japan for inflammatory and autoimmune conditions, with a fascinating profile: *antioxidant, immunomodulatory, membrane-stabilising, and deeply restorative*.

What drew James in was that Cepharanthine didn't just act on immune markers – it worked at the *interface of membrane and signal*, where *cellular communication was governed by lipid fluidity, receptor integrity, and redox balance*.

First, he learned how it suppressed the *NF-κB pathway*, the same inflammatory transcription cascade that had been a common denominator in many of his symptoms. By downregulating NF-κB, Cepharanthine *reduced the expression of IL-4, IL-6, and other Th2-promoting cytokines*, lifting the immune fog that kept his body in reactive overdrive (Bailly, 2019).

But that was only the beginning.

Cepharanthine also activated *AMP-activated protein kinase (AMPK)* – a master regulator of cellular energy. AMPK was not just a metabolism molecule; it was an *immune rheostat*, shifting cellular priorities away from inflammation and towards repair. For James, who had long battled immune exhaustion and metabolic rigidity, AMPK activation felt like a switch flipping. *His energy stabilised. His immune system became less noisy. His mitochondria were more responsive* (Liu et al., 2023).

What stood out most was Cepharanthine's ability to *stabilise plasma membrane fluidity* – a concept that felt abstract until James realised how central it was to immunity. Cell membranes were not just passive borders. They were *interactive platforms*, where antigen receptors clustered, immune messages were transmitted, and cellular decisions were made.

When membrane fluidity is compromised – as it often is in chronic inflammation, lipid peroxidation, and Th2 dominance – *cell signalling becomes chaotic*. Receptors misfire, cytokine responses become exaggerated, and immune cells struggle to coordinate. Cepharanthine, with its amphipathic structure, *reinforced the membrane's integrity*, reducing immune noise and *restoring the precision of intercellular dialogue*.

It was like tuning an orchestra that had been playing out of sync for years.

Cepharanthine's *antioxidant effects* added another layer. By reducing lipid peroxidation and scavenging reactive oxygen species, it eased the *oxidative burden that fed into Th2 skewing*, while also protecting mitochondrial membranes – helping to preserve the T3 thyroid signalling that James had worked so hard to restore (Bailly, 2019).

And perhaps most impressively, Cepharanthine *enhanced an-*

tigen presentation. It modulated the behaviour of B cells and antigen-presenting cells in lymphatic tissues, helping *correct the overactive humoral responses* that had once driven James' food sensitivities, histamine intolerance, and chronic immune activation (Liu et al., 2023).

For James, this was the missing link. A compound that didn't just block inflammation or stimulate immunity, it *reorganised the immune terrain at the microstructural level*.

He noticed it not as a burst of energy, but as *a quiet resilience*. He had fewer setbacks after exposure to stress, less reactivity to food and weather, and a steadier mind. A body no longer needed to speak in symptoms.

Cepharanthine wasn't loud. It was *foundational*. It worked where medicine often doesn't look – *at the membrane*, at the *architecture of immunity*, at the *space between signals*.

And through that, James realised something vital: sometimes healing doesn't mean doing more. It means *stabilising the platform* so that everything else – nutrition, hormones, immune signals, even thoughts – can integrate.

This wasn't suppression. It was *clarity through structure*.

And that structure, finally restored, was becoming the *true baseline of health*.

BEFORE IMMUNE RETRAINING
Th2-BIASED PATTERNS
T CELL
T CELL
B CELL
B CELL
AFTER IMMUNE RETRAINING
BALANCED PATTERNS
B CELL
T CELL
B CELL
REGULATION
WELLNESS
PROTECTION
ADAPTIVE IMMUNITY
LESS INFLAMMATION

Chapter 26

Whispered Signals: Micro-Immunotherapy and the Rewriting of Th2 Memory

By this point in his journey, James understood that not all immune interventions needed to be forceful. Some of the most transformative signals were those that *whispered* and aligned with the body's own language of regulation and repair.

He had corrected the metabolic, structural, and hormonal drivers of his Th2 dominance, restoring mitochondrial balance and calming cytokine storms.

But the final piece felt more subtle – *how to retrain the immune system's memory without overstimulation*? How to nudge it into clarity rather than command it into silence?

That is when James encountered *micro-immunotherapy*.

At first, it seemed almost too gentle – *nanodoses of immune messengers*, designed to modulate immune function without triggering resistance or rebound. But as he explored further, he came to see it not as weak medicine, but as *precision signalling*, leveraging the intelligence of *low-dose cytokines*, *interleukins*, and *immune regulators* to *recalibrate from within*.

He was introduced to three core formulations, each targeting a distinct facet of his lingering immune imbalance: *ALERG*, *INFLAM*, and *DEP*.

ALERG

This formula focused on the *core architecture of Th2 dominance* – downregulating *IL-4, IL-5, IL-10, and IL-13*, the same cytokines that had once driven his eczema, hay fever, and post-meal histamine flushes. ALERG also targeted *IgE production and mast cell activation*, reducing the *overreactive humoral responses* that had made his immune system fire too easily, too often.

Within weeks of use, he noticed *less skin reactivity*, *fewer respiratory symptoms*, and a *greater tolerance to previously triggering foods and environments*. ALERG wasn't downregulating his immune system – it was simply *reminding it of balance*.

INFLAM

Next came INFLAM – a formulation designed not just to reduce inflammation, but to *resolve it*. Through the *downregulation of IL-6 and TNF-α*, and the gentle upregulation of *IL-10 and TGF-β*, INFLAM supported a *shift away from smouldering systemic inflammation* and towards *tissue repair and immune recalibration*.

James felt this not only in reduced muscle tightness and joint aches, but in *his emotional bandwidth*. The inflammation that once clouded his clarity and shortened his fuse was *quieting*, creating space for thought, presence, and recovery.

DEP

The third formula, DEP, surprised him the most. It didn't just modulate immune responses – it addressed *neuroimmune balance*, the space where *cytokines meet cognition*, where inflammation blurs into fatigue and fog. DEP gently adjusted *IL-1, IL-2, IL-4, IL-6*, and *TGF-β*, reducing *neuroinflammation* and stabilising the *cross-talk between his immune system and nervous system*.

This was where James felt *his sleep deepen*, his dreams return, and his *sense of agency over his body reawaken*. DEP gave his brain a chance to rest – not through sedation, but through *resolution*.

Micro-immunotherapy didn't work overnight. It wasn't designed to. But James came to appreciate that *biology doesn't always need a hammer*. Sometimes it just needs a *nudge in the right direction, delivered in the language it remembers best*.

He also appreciated the *targeted nature* of the therapy. Each formula worked with specificity – *not a broad immune downregulator, but a rebalancer*. A low-dose, high-precision approach that *tuned rather than muted*.

The therapeutic implications became clear:

- In ***allergic conditions,*** ALERG helped reduce hyper-reactivity while preserving immune vigilance.
- In ***autoimmunity and chronic inflammation,*** INFLAM restored cytokine ratios and improved tissue tolerance.
- In ***neuroinflammatory syndromes,*** DEP offered a calm reset for a brain burdened by years of over-signalling.

For James, micro-immunotherapy wasn't a supplement or a medication. It was a *bridge* – between immune chaos and immune coherence, between overreaction and restored recognition.

He no longer feared his immune system. He was finally *in relationship with it*.

And through micro-immunotherapy, he had learned something profound: *Healing is not just about what you suppress – it is about what you teach*.

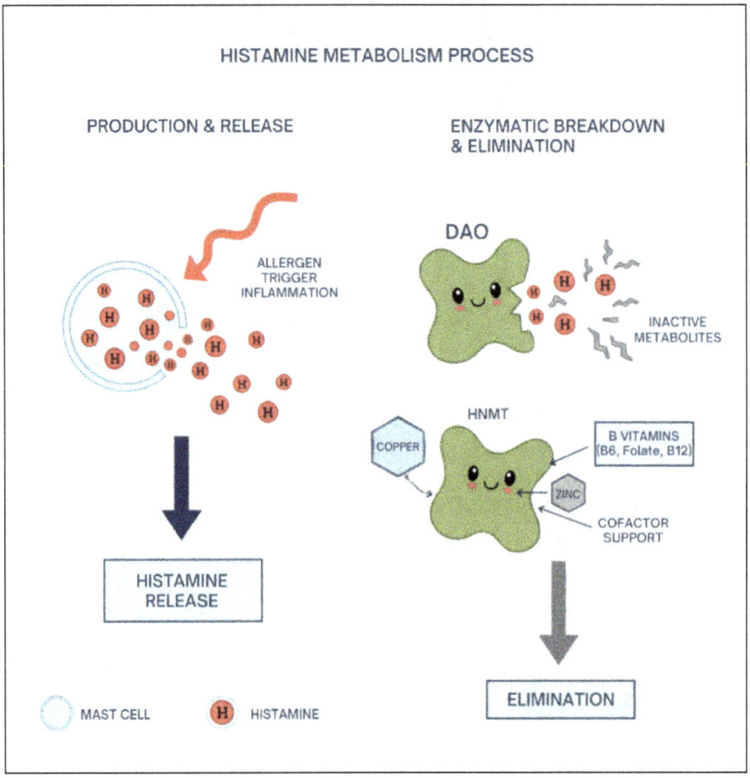
HISTAMINE METABOLISM PROCESS
PRODUCTION & RELEASE
ENZYMATIC BREAKDOWN & ELIMINATION
ALLERGEN TRIGGER INFLAMMATION
DAO
INACTIVE METABOLITES
HNMT
COPPER
B VITAMINS (B6, Folate, B12)
ZINC
COFACTOR SUPPORT
HISTAMINE RELEASE
ELIMINATION
MAST CELL
HISTAMINE

Chapter 27

Modulating the Signal: Histamine, Th2 Dominance, and Nutritional Precision

By now, James had untangled many threads in his immune story. Yet one theme had quietly run through nearly every chapter: *histamine*.

It had shown up in *skin flushes* after meals. In *sinus pressure* and *itchy eyes* when seasons changed. In the *waking anxiety* and *tight chest* that followed histamine-liberating foods or stressful days. Histamine was not the enemy – it was a signal. But in his Th2-skewed terrain, the signal had become *distorted, excessive, and chronically unresolved.*

As James came to understand, *histamine was not just an allergic molecule – it was an immune modulator*, a neurotransmitter, and a regulator of vasodilation, acid secretion, and alertness. But in a Th2-dominant state, where *mast cells were overactive* and *IL-4, IL-5, and IL-13* drove chronic inflammation, histamine became a fuel source for reactivity.

The solution wasn't to suppress it outright. It was to *modulate* it – to address the enzymes that broke it down, the cells that released it, and the oxidative terrain that amplified its effects.

He began with *quercetin* – a flavonoid with profound mast cell stabilising properties. It worked upstream, not just by *preventing degranulation*, but by *inhibiting histidine decarboxylase*, the enzyme that converted histidine into histamine. Quercetin gave

James *space* – reducing his reactivity to allergens and improving his tolerance to histamine-rich foods. It wasn't just symptom relief – it was *modulating the volume of the signal*.

Then came *vitamin C*, which surprised him with its subtlety. Beyond its antioxidant power, vitamin C *increased diamine oxidase (DAO)* – the enzyme responsible for degrading histamine in the gut. It also stabilised mast cells and helped recycle glutathione. But James was cautious: high doses had been linked to *increased IL-4 production*, potentially exacerbating Th2 dominance in some contexts. He found that *500–1,000 mg/day* gave him the benefit without overstimulation.

Zinc became another cornerstone – *supporting DAO function*, reducing histamine receptor sensitivity, and aiding epithelial repair. It worked quietly, but deeply – particularly in *skin health* and *gut barrier restoration*. With zinc, James saw improvements not only in immune tone but in *immune boundaries*.

Then there was *N-acetylcysteine (NAC)* – a compound James had already used for glutathione support. Now he understood its additional role in *mast cell stabilisation* and *histamine metabolism*. It was particularly helpful in clearing *chronic sinus congestion* and improving *respiratory resilience*. But, again, nuance mattered: doses above *2,000 mg/day* could *inhibit redox-sensitive immune pathways*, delaying resolution. He found his balance around *600–1,200 mg/day*, depending on need.

His journey into the gut brought him to *probiotics*. At first, he assumed all probiotics were equal, but he quickly discovered that certain strains – especially some *Lactobacillus* species – could *stimulate Th2 cytokines like IL-4 and IL-10*, exacerbating the very patterns he was trying to resolve. He shifted instead to *Bifidobacterium longum* and *Lactobacillus plantarum*, strains shown to support *Th1 balance*, reduce *histamine-producing bacteria*, and enhance *DAO activity*. The result? *Better digestion, fewer rashes, clearer mornings.*

Still, there were moments where food reactions crept in –

when aged cheese or fermented foods tipped him over the edge. That is when *DAO supplementation* came into play. Taken just before meals, it helped *degrade dietary histamine* in the gut. But James knew its limitations. DAO wasn't a fix – it was a *bridge*. Its effectiveness depended on gut health, zinc availability, and *whether the permeability and inflammation that drove his histamine load were being addressed at their source*.

He paired DAO with *gut-healing nutrients* – zinc carnosine, glutamine, and short-chain fatty acids – to support *long-term restoration of mucosal integrity*. Over time, his reliance on DAO lessened as his gut environment became less reactive, less porous, and more *intelligent*.

Through this phase, James learned an essential truth: *histamine was not inherently bad*. It was a *context-dependent messenger*, and in the right environment it played crucial roles in vigilance, immunity, and digestion. The issue wasn't histamine itself – it was the *terrain* that allowed it to become excessive, toxic, or trapped.

By modulating that terrain with nutrients, enzymes, and precise probiotics, James didn't suppress histamine – he *restored its wisdom*.

He moved forward with a body that no longer braced at every trigger. One that could *respond, not overreact*. And perhaps most importantly, a body that trusted its own signals again.

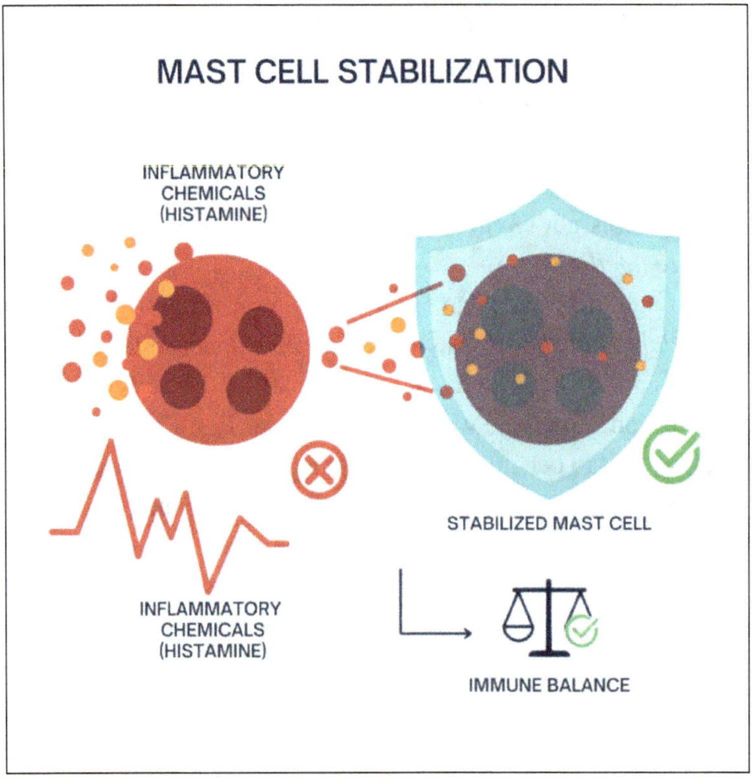

MAST CELL STABILIZATION
INFLAMMATORY
CHEMICALS
(HISTAMINE)
STABILIZED MAST CELL
INFLAMMATORY
CHEMICALS
(HISTAMINE)
IMMUNE BALANCE

Chapter 28

The Gentle Guard: Mast Cell Stabilisers and the Containment of Immune Overreaction

James had learned to listen not just to the loud flares – rashes, congestion, crashes – but to the *subtler signs*. A flutter of skin heat. A low, dull headache after a histamine-rich meal, or the tightness in his chest after unexpected stress. These were not dramatic alarms – they were the *final echoes of immune volatility*, a whisper from *mast cells*, ever on the brink of degranulation.

He had come to understand that mast cells were not rogue agents – they were ancient, fast-responding sentinels. But in the Th2-dominant landscape his body had long inhabited, their thresholds had been *lowered*, their signals *distorted*.

What James needed now wasn't more immune activation – it was *containment*. A way to *stabilise mast cells*, reduce unnecessary histamine release, and restore the boundaries between *internal environment and external stimuli*.

And so, he turned to a class of allies he had previously overlooked: *natural mast cell stabilisers*.

Luteolin

The first was *luteolin*, a potent flavonoid with dual action: it *blocked mast cell degranulation* and *inhibited NF-κB,* the master switch of inflammation. Within days of taking it, James noticed his skin had felt less reactive, his breath deeper. But he remained cautious – *luteolin had mild anticoagulant effects*, and though he wasn't on blood thinners, he respected its power. At doses of *100–300 mg/day*, it became a *precision tool* – small, sharp, and deeply effective.

Curcumin

Then came *curcumin* – an old friend in his protocol, now reintroduced with *renewed purpose*. Beyond its systemic anti-inflammatory action, curcumin was a *mast cell modulator*, reducing IL-6, TNF-α, and the downstream release of histamine. But James knew its limitations. The standard form being barely absorbed, he shifted to *liposomal curcumin*, paired with *black pepper extract (piperine)* to enhance bioavailability. He stayed within *500–1,000 mg/day*, watching carefully for signs of bile intolerance, given his history of sluggish gallbladder function.

Magnesium

Magnesium had already played a quiet role in his recovery, calming his nervous system and aiding sleep. Now, he understood it *modulated calcium signalling* at the level of mast cells, reducing the intracellular triggers for degranulation. At doses of *200–400 mg/day*, especially in *glycinate or citrate forms*, magnesium brought a stabilising effect to his physiology, and to his sense of *internal steadiness*.

He also noticed that when his magnesium was optimal, *gut symptoms improved,* with less bloating, fewer cramps, and smoother digestion. The systemic effect of this single mineral reminded him just how *interconnected every system truly was*.

Resveratrol

Resveratrol surprised him. Known for its antioxidant and anti-ageing properties, it also had a *powerful mast-cell stabilising effect*, largely through reducing *oxidative stress and inflammatory cascades*. But it came with a caveat: *oestrogenic potential*. Resveratrol mimicked oestrogen at certain receptors, and in individuals like James – prone to oestrogen dominance – *dosing had to be cautious*.

He settled on *200–400 mg/day*, while monitoring his mood, energy, and hormonal rhythms. In James' case, it worked best as a *short-term anti-inflammatory*, when histamine load crept up.

Bromelain

Finally, there was *Bromelain* – an enzyme from pineapple, long known for its role in protein digestion. What James hadn't known was that Bromelain also helped *reduce mast cell activity*, in part by lowering the *antigenic burden* of incompletely digested proteins. It was not only reducing histamine reactivity – it was also *reducing the stimuli that triggered it*.

He used *500–1,000 mg/day*, carefully avoiding it on days when he had eaten raw pineapple (a rare occurrence, but a known trigger). In combination with a *low-histamine diet and digestive enzymes*, Bromelain became an essential tool in *managing the threshold of immune tolerance*.

As he integrated these stabilisers, James noticed something different – not just fewer symptoms, but *a deeper resilience*. His system no longer buckled under minor exposures. His energy held steady in crowded rooms, during seasonal changes, and even after emotional stress.

The mast cells, once quick to react, now *stood their ground with greater discernment*.

James had come to see mast cells not as defective, but as *exquisitely tuned receptors*, whose job was to sense and protect. His goal had never been to silence them – but to *help them remember the difference between threat and signal*.

With the right nutrients, carefully chosen and dosed, he was no longer managing an immune system on edge. He was restoring *an immune system in relationship* – with the environment, with the microbiome, and with his nervous system.

These stabilisers weren't suppressing his system.

They were *reinforcing its integrity*.

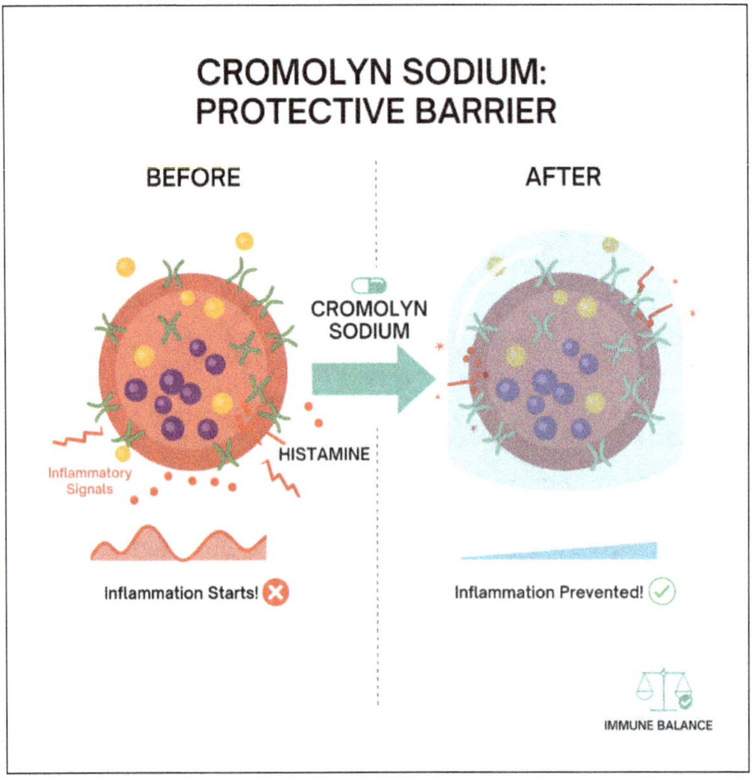
CROMOLYN SODIUM:
PROTECTIVE BARRIER
BEFORE
AFTER
CROMOLYN
SODIUM
HISTAMINE
Inflammatory
Signals
Inflammation Starts!
Inflammation Prevented!
IMMUNE BALANCE

Chapter 29

Holding the Line: Cromolyn Sodium and the Quieting of Th2 Fire

James had learned to think of his immune system not as a battlefield, but as *a network of borders* – barriers that protected, communicated, and adapted. And nowhere were these borders more strained than at his *epithelial surfaces*: the lungs, the gut, the skin. These were the *places where the inside met the outside*, where *mast cells gathered like sentinels*, ready to respond.

And for years, they had been *over-responding*.

Despite all he had done – modulating histamine, supporting detox, rebalancing mitochondria – there were still moments when his immune system acted *before thinking*, releasing histamine and inflammatory mediators at the slightest perceived insult. James wasn't at war anymore – but he still lived in a body that flinched.

That's when he revisited *cromolyn sodium*.

Once used only in his asthma puffer during high school track meets, cromolyn was something James had nearly forgotten. But now, through a different lens, he saw it not as a mere allergy drug, but as *a deeply strategic mast cell stabiliser*, capable of *disrupting the loop of Th2 dominance* at its foundation.

Cromolyn worked by *stabilising mast cell membranes*, preventing the release of *histamine*, *leukotrienes*, and *cytokines* like IL-4, IL-5, and IL-13. These were the *master signals* that perpetuated Th2 dominance – recruiting eosinophils, driving IgE production, and maintaining the cycle of inflammation that had dominated James' immune narrative for years.

The difference now was that James saw cromolyn not as suppression – but as *containment.*

It didn't override immune function. It simply *held the threshold*, buying time for recalibration.

One of its most powerful effects came through the *inhibition of eosinophilic inflammation*. Eosinophils had played a destructive role in James' past – driving his allergic asthma, contributing to gut inflammation, even affecting his skin during seasonal transitions. With cromolyn, he felt a *softening* in these tissues. Less tightness in the chest. Less itching. Less need for the body to scream through symptoms.

But cromolyn's effects went deeper still – into the *mucosal immune system*, where much of the Th2-skewed activation began. In the gut, cromolyn *stabilised intestinal mast cells*, reducing histamine release, preserving barrier integrity, and protecting him from the cascade of reactions that once followed meals or stress. In the sinuses and airways, it reduced hyper-reactivity, *allowing the epithelial lining to heal* rather than perpetually defend.

And importantly, cromolyn *normalised histamine signalling* – not just reducing its release but preventing it from *amplifying Th2 cytokine production via dendritic cell activation*. This subtle interaction had kept James stuck in feedback loops, where each histamine surge reinforced the very immune pathways he was trying to escape.

Now, with cromolyn on board, those loops were *interrupted.*

The therapeutic implications were clear:

- In ***allergic conditions*** like asthma, hay fever, and eczema, cromolyn *lowered the frequency and intensity of flares*.
- In ***chronic inflammatory states***, especially where eosinophils were dominant, cromolyn *reduced tissue burden and inflammatory signalling*.

- In ***gut disorders***, cromolyn supported the restoration of *mucosal integrity*, helping to rebuild an environment where the immune system could engage *without overreacting*.

James did not need to use it in the long term, as it wasn't about dependency. It was about *a window of safety* – giving his immune system time to unlearn its reactivity, to reclaim tolerance, and to *hold the line without flinching*.

This chapter of his journey taught him that healing wasn't always loud. Sometimes, it was found in the *stillness between triggers*. In the *quiet that follows containment*.

Cromolyn sodium had given his immune system the chance to *pause*, to *respond rather than react*. And in that pause, James found a space he hadn't felt in years:

Peace at the threshold.

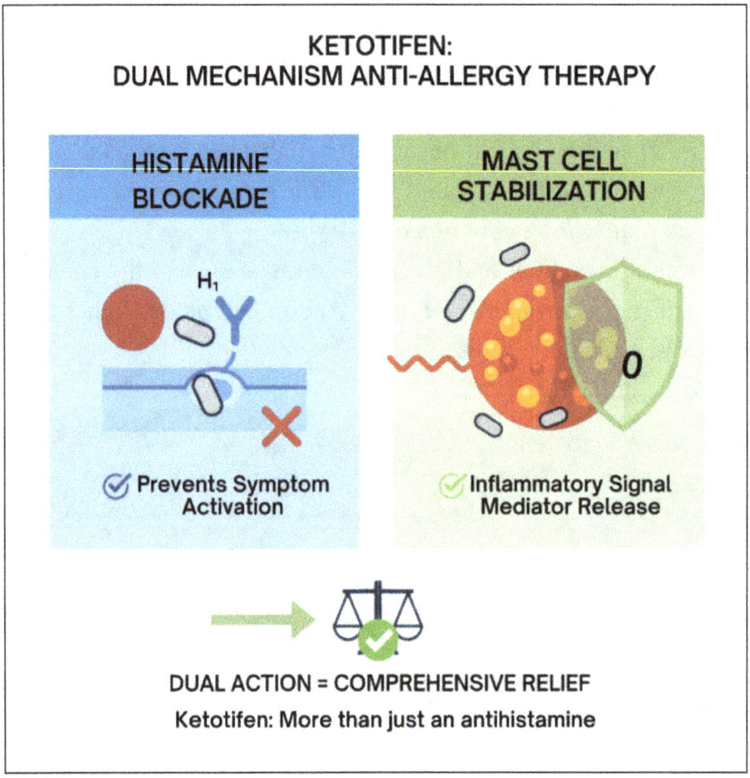
KETOTIFEN:
DUAL MECHANISM ANTI-ALLERGY THERAPY
HISTAMINE BLOCKADE
H_1
Prevents Symptom Activation
MAST CELL STABILIZATION
0
Inflammatory Signal Mediator Release
DUAL ACTION = COMPREHENSIVE RELIEF
Ketotifen: More than just an antihistamine

Chapter 30

Two Doors, One Key: Ketotifen and the Precision Modulation of Th2 Dominance

The immune system doesn't always need a hammer – it needs a *key*. A way in. A mechanism not just to suppress, but to *redirect*, to relieve, and to resolve.

After finding stability through cromolyn sodium, which helped *prevent mast cell degranulation*, James had gained a new-found sense of quiet. But in moments where his immune system still flared – such as after a restaurant meal, during pollen spikes, or on days when his body felt inexplicably 'on edge' – he realised he needed *a tool that could do more than prevent activation. He needed one that could also buffer the signals already in motion.*

That tool arrived in the form of *ketotifen.*

Unlike cromolyn, ketotifen offered *dual action*: a *mast cell stabiliser* and a *histamine H1 receptor antagonist*. It didn't just stop mast cells from releasing inflammatory mediators – it also *blocked histamine's downstream effects* once it had been released. For someone with a long-standing Th2 dominance, this was critical. Ketotifen *intervened both at the source and at the signal.*

At the mast cell level, ketotifen inhibited *degranulation*, just like cromolyn. But it also downregulated the release of *IL-4, IL-5, and leukotrienes* – the drivers of *eosinophilic inflammation* and *IgE production* that had once ruled James' allergic reactivity.

Where cromolyn required consistent, pre-emptive use, *ketotifen offered immediate relief*, especially for *symptoms already in motion*: the flare of urticaria, the onset of sinus congestion, or the sudden wave of histamine-induced fatigue.

But what James appreciated most was its impact on *eosinophilic activity*. For years, eosinophils had lingered in his tissues, waiting for any opportunity to inflame his gut lining, lungs, and skin. Ketotifen *interrupted their recruitment and activation*, not just calming symptoms, but *stopping tissue-level inflammation before it took root.*

And unlike many other antihistamines, ketotifen didn't act only peripherally – it had a *deep immunomodulatory effect*, helping shift the *Th2-dominant immune pattern towards greater balance*.

But ketotifen, like all medicines, required *precision*.

James soon learned that it was metabolised through the *CYP3A4 enzyme*, part of the cytochrome P450 system in the liver. Genetic testing had already revealed that he was a *slow metaboliser*, which explained why other medications often lingered longer in his system. With ketotifen, this meant *a longer duration of action*, but also a *greater sensitivity to sedative effects*, especially when taken in the evening.

He adjusted accordingly – *lower doses*, taken at the right time, with an eye towards the balance between relief and clarity.

In this way, ketotifen became both *a bridge and a buffer*. It gave James more room between stimulus and response. Between histamine release and full-body reaction. Between trigger and flare.

He no longer feared the edge. Because now, he had *a margin*.

The therapeutic implications became clear:

- In ***allergic conditions***, ketotifen offered both *fast relief* and *long-term recalibration*.
- In ***chronic inflammatory disorders***, it reduced *eosinophilic burden*, helping restore immune boundaries in the lungs, gut, and skin.

- In ***personalised care***, it required thoughtful dosing, particularly for those with *liver enzyme polymorphisms* – not as a limitation, but as an opportunity for *tailored therapy*.

For James, ketotifen was not about reliance – it was about *empowerment*. A way to live more freely, with fewer braced muscles and fewer surprises from a once-unpredictable immune system.

This wasn't sedation. This was *precision immunology*, paired with personal awareness.

Another tool. Another step.

And with each, James was not just recovering from Th2 dominance – he was *reclaiming his capacity to respond, not react*.

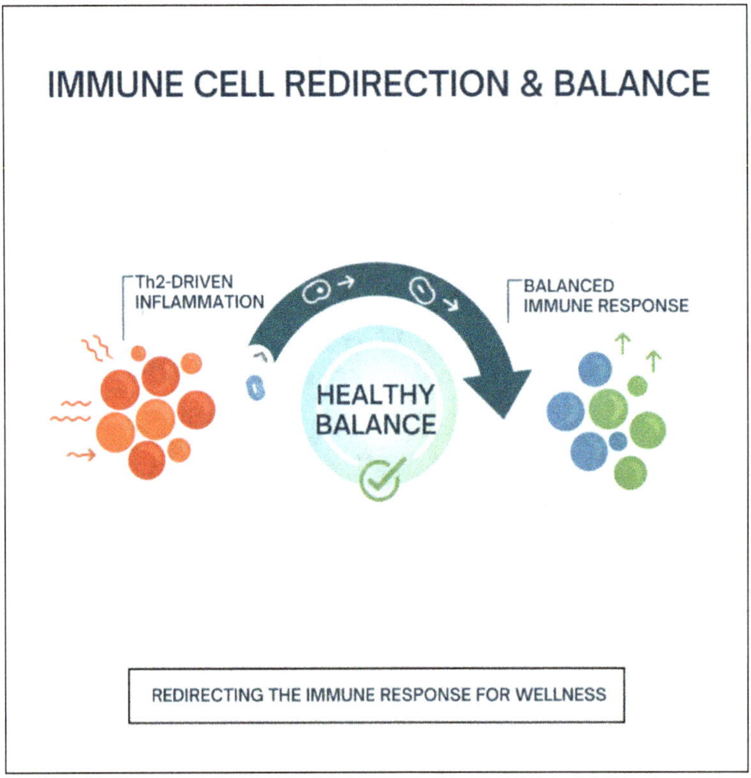
IMMUNE CELL REDIRECTION & BALANCE
Th2-DRIVEN INFLAMMATION
HEALTHY BALANCE
BALANCED IMMUNE RESPONSE
REDIRECTING THE IMMUNE RESPONSE FOR WELLNESS

Chapter 31

Changing the Direction of the Signal: Maraviroc and the Chemokine Correction of Th2 Dominance

James had come to view his immune system as a conversation in motion – not just between cells, but across tissues, across time, across internal landscapes. He now understood that how immune cells migrated – *where they went, what they spoke to, and when they arrived* – mattered just as much as what they secreted.

And in the landscape of chronic Th2 dominance, *the traffic had been misrouted for years.*

While many of his symptoms had softened – thanks to mast cell stabilisers, histamine modulation, and mitochondrial repair – James still experienced episodes of immune overactivation, particularly at *epithelial interfaces*. Skin flares. Sinus congestion. Fatigue after meals. It was clear: the deeper circuitry of his immune response still needed some rewiring.

That is when he discovered *Maraviroc*, a CCR5 antagonist with a unique ability to influence *immune cell migration and signalling.* Originally developed as an antiretroviral, its broader application lay in *blocking chemokine-driven pathways*, particularly those that regulated *Th2 cell recruitment and activation* (Abel et al., 2008; Abel et al., 2009).

By binding to *CCR5 receptors*, Maraviroc interfered with the *chemotactic signals* that drew Th2 cells to tissues. This disruption helped *downregulate the production of IL-4, IL-5, and IL-13* – cytokines that had long sustained his allergies, eosinophilic activity, and chronic inflammation.

But what made Maraviroc exceptional was not its suppressive effect – it was its ability to repattern cellular movement, restoring immune equilibrium without inducing inhibition.

James noticed the changes in his *airways* and *gut lining* – tissues once prone to flare at the slightest insult. Maraviroc reduced *eosinophil recruitment* and *basophil activation*, easing the chronic inflammatory load and calming the immune overreaction that had once required constant vigilance. The inflammation didn't just subside – it *began to resolve*.

There was also an impact on *Treg activity* – not through direct stimulation, but by *reducing the inflammatory environment* that had previously suppressed their expression. With cytokine load diminished and tissue signalling stabilised, *regulatory pathways re-emerged*, promoting a more *tolerant, nuanced immune response*.

This was not about flipping the immune system from one side to another. It was about *modulating its tone* – reducing hyper-reactivity and allowing *adaptive intelligence* to return.

Still, James approached with care. Maraviroc was metabolised through *CYP3A4*, and his slow liver detox capacity meant that it had a *longer half-life* in his system. Working with his health practitioner, he started with a low dosage and adjusted it carefully to avoid side effects like *light-headedness or postural shifts*. With personalised dosing, it became a *precise, quiet intervention* – subtle, but powerfully stabilising.

He appreciated Maraviroc's uniqueness. Unlike corticosteroids or biologics that often came with trade-offs, Maraviroc *redirected immune traffic without suppressing function*. It allowed his body to choose *when to respond, where to respond, and how much was necessary*.

The implications were significant:

- In ***Th2-mediated disorders***, it helped *attenuate cytokine signalling and tissue recruitment*, reducing chronic inflammation without dulling defence.
- In ***eosinophil-driven conditions***, it lowered *tissue damage and fibrosis*, supporting repair without creating vulnerability.
- In ***immune modulation***, it offered *tailored rebalancing*, not by forcing direction, but by *clearing space for the immune system to regulate itself.*

For James, this was not just another therapy. It was a lesson in *interception*. In the power of changing the *movement* of the immune system, not just the molecules which it produces.

And in that, he saw the wisdom he had been moving towards all along: *Immune health isn't built by dominating responses – it is built by restoring options.*

With Maraviroc, his body could finally *choose again*.

And choice, James realised, was the foundation of resilience.

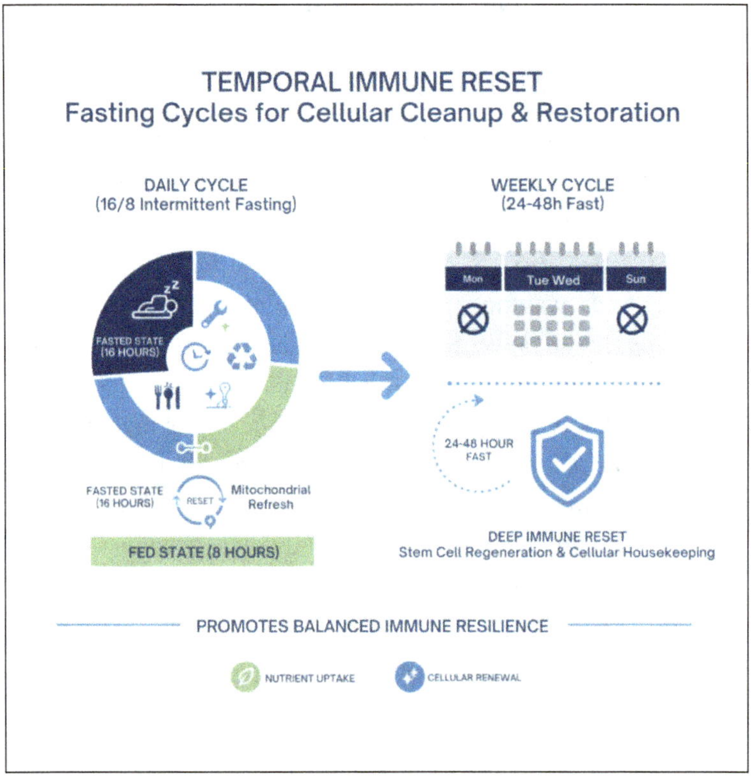
TEMPORAL IMMUNE RESET
Fasting Cycles for Cellular Cleanup & Restoration
DAILY CYCLE
(16/8 Intermittent Fasting)
WEEKLY CYCLE
(24-48h Fast)
FASTED STATE
(16 HOURS)
Mon
Tue Wed
Sun
24-48 HOUR
FAST
FASTED STATE
(16 HOURS)
RESET
Mitochondrial
Refresh
FED STATE (8 HOURS)
DEEP IMMUNE RESET
Stem Cell Regeneration & Cellular Housekeeping
PROMOTES BALANCED IMMUNE RESILIENCE
NUTRIENT UPTAKE
CELLULAR RENEWAL

Chapter 32

The Rest Between Signals: Fasting and the Repatterning of Th2 Dominance

James had spent years learning how to nourish his system – through targeted nutrients, immunomodulators, mast cell stabilisers, and neuroimmune therapies. But as his physiology became more stable, he began to sense a different kind of need. Not for more, but for *less*. A quieter recalibration. A kind of *physiological stillness*.

That stillness came in the form of *fasting*.

It wasn't new to him in concept. He had experimented with intermittent fasting before – mostly for focus or metabolism – but now he understood it as *a tool for immune modulation*, one that spoke the language of his body in a new, deeper way.

He came to see fasting not as deprivation, but as *repatterning* – a strategic withdrawal of input that allowed his system to reset inflammation, re-engage autophagy, and re-establish tolerance.

The science made sense. Fasting lowered levels of *IL-6 and TNFα*, key cytokines that had sustained his Th2-dominant profile. As those inflammatory signals dropped, the environment became more conducive to *immune rebalancing*, not through suppression, but through space – *the absence of perpetual activation*.

Autophagy became one of the most powerful shifts. As James extended his fasts to 14, sometimes 16, hours, he noticed improvements in *mental clarity, digestion, and immune calm*. Autophagy –

his cells' internal cleanup crew – was removing *damaged mitochondria, residual pathogens, and lingering inflammatory debris*. It was clearing the stage for something new to emerge: *homeostasis*.

Even more remarkable was the role of *ketosis*. During longer fasting windows, James entered a mild ketogenic state, marked by rising levels of β*hydroxybutyrate (BHB)*. Far from just a fuel source, BHB suppressed the *NLRP3 inflammasome*, a key driver of innate immune overactivation and Th2 amplification. This, too, was part of the quieting.

His gut responded in kind. Fasting reshaped his microbiome – not only by resting his digestion, but by shifting microbial composition. Pathogenic and histamine-producing bacteria declined, while beneficial strains increased. His histamine symptoms lessened, and his digestion became more *efficient and less reactive*.

Even *mast cell activity*, which had once spiked at every unknown food or social stressor, began to soften. With reduced energy substrates, mast cells became *less prone to degranulate*, and circulating *IgE levels dropped* over time. It was as if fasting helped *teach his immune system restraint* – to pause before reacting.

But James knew fasting was not without risks. With his history of *HPA axis dysregulation*, extended fasting could backfire – elevating cortisol, suppressing TSH, or exacerbating adrenal fatigue. He kept his fasts between *12 and 16 hours*, ensuring his nervous system remained *in parasympathetic tone*, not survival mode.

When it came to thyroid health, he treaded carefully. He knew that *fasting reduced peripheral T3* – not normally problematic, but in someone with Th2-driven thyroid resistance, it could lead to *hypothyroid-like symptoms*. He decided to balance fasting with strategic *refeeding windows* rich in *selenium, zinc, iodine, and tyrosine* – supporting hormone conversion, mitochondrial output, and metabolic recovery.

Over time, he developed a rhythm:

- ***Intermittent fasting*** on low-demand days.
- ***Shorter eating windows*** with high-nutrient meals.
- ***Planned rest days*** when adrenal output was low.
- ***Mindfulness and breathwork*** to buffer cortisol spikes during longer fasting periods.

And with this rhythm, James found something extraordinary.

Not just immune improvement, but *neuroimmune coherence.*

His brain fog dissipated. His circadian rhythm deepened. His inflammatory markers stabilised. He no longer feared food, or the gaps between meals. Instead, he used those gaps to listen – to reset – to let *the immune system breathe.*

Fasting had become a space where the body *relearned rhythm, tolerance, and trust.*

It wasn't about pushing. It was about *pausing* – so the immune system could step out of the pattern of constant readiness, and into *adaptive intelligence.*

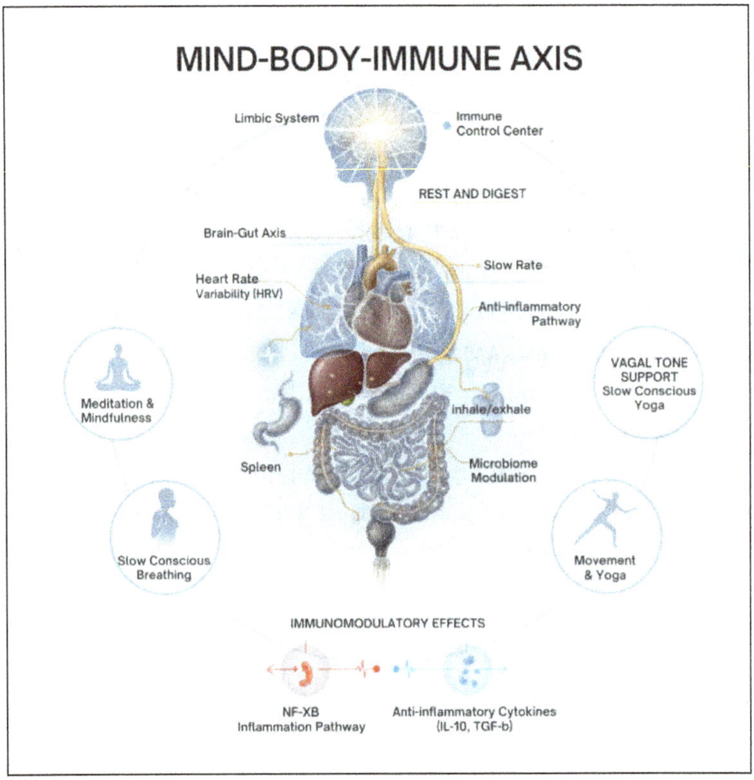
MIND-BODY-IMMUNE AXIS
Limbic System
Immune Control Center
REST AND DIGEST
Brain-Gut Axis
Slow Rate
Heart Rate Variability (HRV)
Anti-inflammatory Pathway
VAGAL TONE SUPPORT Slow Conscious Yoga
Meditation & Mindfulness
inhale/exhale
Spleen
Microbiome Modulation
Slow Conscious Breathing
Movement & Yoga
IMMUNOMODULATORY EFFECTS
NF-XB Inflammation Pathway
Anti-inflammatory Cytokines (IL-10, TGF-b)

Chapter 33

Returning Home: Vagal Tone, Limbic Repair, and the Embodiment of Immune Resilience

James had addressed almost every system – thyroid and mitochondria, mast cells and histamine, cytokines and chemokines. He had rewired his immune architecture piece by piece. But as the dust settled, a deeper layer emerged: *the nervous system.*

It had been there the entire time, humming beneath every flare, every food reaction, every sleepless night. The skin that flushed after stress. The breath that tightened during confrontation. The gut that turned to fire with the wrong glance or the wrong food. These were not just immune responses.

They were *neuroimmune echoes*.

James realised that while much of his immune restoration had focused on *inputs and molecules*, the true conductor of the entire system – the one which decided whether the body was in defence or repair – was *the limbic brain*. And seated at the core of that decision was the *vagus nerve*.

For years, James had lived in a subtle state of threat – a kind of unconscious hypervigilance. His nervous system had been trained by chronic illness to expect reactivity. That expectation had become *a physiological loop*, where even after inflammation resolved, his brain still braced for it.

This was where *limbic retraining* began.

It was not about rewiring belief. It was about *teaching the nervous system that safety was possible again.*

He began with gentle practices – *neuroplastic repetition*, visualisations, and *emotional re-association.* He used positive affirmations, scents, and daily calming rituals to re-link neutral or positive experiences in physiological states that were previously associated with pain. Slowly, the nervous system *disengaged from its pattern of overreaction.*

Then came the *vagus nerve* – the superhighway of rest, digestion, and immunity. James learned how vagal tone determined his immune setpoint. When vagal tone was strong, *inflammation calmed*, digestion optimised, and Treg activity rose. When the tone was weak, *inflammatory loops returned*, regardless of how 'clean' the rest of his system was.

So, he turned to *vagal toning practices*:

- ***Slow, coherent breathing*** – in through the nose, out longer than the inhale.
- ***Cold exposure***, ending his showers with brief bursts of icy water.
- ***Humming, chanting, and gargling***, activating vagal branches through the voice and throat.
- ***Gentle movement***, like ***yoga and somatic shaking***, to discharge trapped stress from his fascia and vagal sheath.

But perhaps the most powerful shift came through *interoception* – learning to feel again, not just think. He practised sensing his heartbeat, noticing tension, naming emotional states. He invited his body back into conversation – not as a battlefield, but as a home.

And it responded.

His startle reflex softened. His breath deepened. His sleep became a place of repair rather than survival. Even his immune markers shifted subtly – *lower IL-6, improved Treg ratios, and less cytokine variability* after stress.

James was no longer just recovering from Th2 dominance.

He was *reclaiming a body that could hold peace without flinching*.

This chapter of vagal restoration taught him that healing didn't just happen through suppression or supplementation. It came through *presence*. Through the courage to sit with a body once filled with chaos – and offer it coherence.

Immune resilience, he now understood, was not the absence of reactivity. It was the ability to return to centre after the storm.

And with every breath, every micro-shift in his nervous system, James was learning to do just that:

Return.

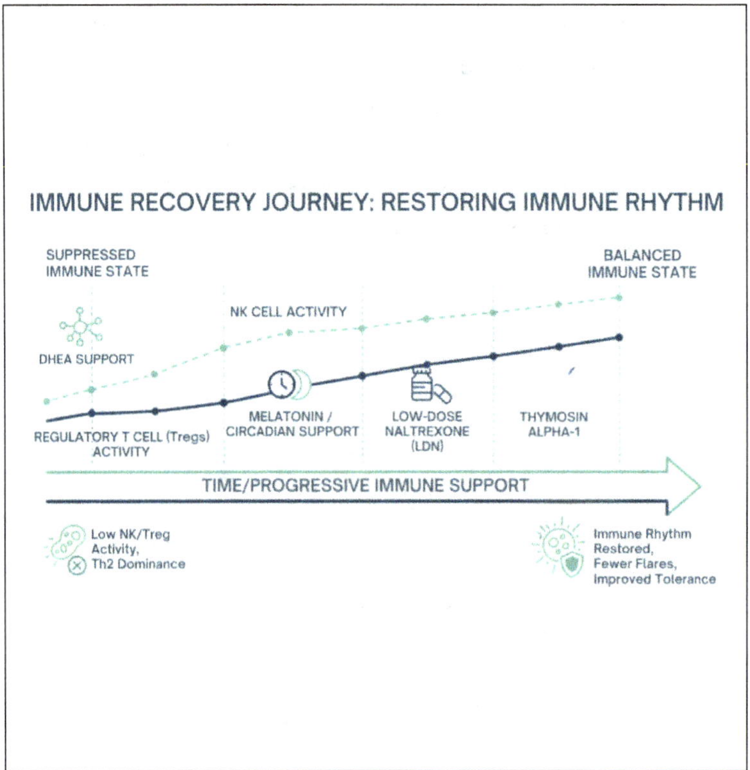
IMMUNE RECOVERY JOURNEY: RESTORING IMMUNE RHYTHM
SUPPRESSED IMMUNE STATE
BALANCED IMMUNE STATE
NK CELL ACTIVITY
DHEA SUPPORT
MELATONIN / CIRCADIAN SUPPORT
LOW-DOSE NALTREXONE (LDN)
THYMOSIN ALPHA-1
REGULATORY T CELL (Tregs) ACTIVITY
TIME/PROGRESSIVE IMMUNE SUPPORT
Low NK/Treg Activity, Th2 Dominance
Immune Rhythm Restored, Fewer Flares, Improved Tolerance

Chapter 34

Reading the Map: Lymphocyte Typing and the Tracking of Th2 Dominance

James had travelled far in his healing – not just across symptoms, supplements, or diagnostic labels, but through the terrain of his own immune system. And while he now lived with a sense of balance he never thought possible, he knew that sustained wellness wasn't static. It was adaptive. Informed. Alive. Wellness was not a fixed state to be reached, but a rhythm to be listened to, a set of internal patterns to be understood over time.

It was time to track that adaptability. To listen to what the cells were saying, not just what the body was feeling. Lymphocyte typing – a tool he once would have dismissed as too technical, too clinical, too far from the felt experience – had become deeply relevant. With his deeper understanding of Th2 dominance, immune architecture, and cytokine-driven inflammation, it now felt like the next step: the ability to see the internal immune terrain in real time, to trace the contours of a pattern that once ruled him invisibly. It was the first time his immune system had spoken in numbers, not just symptoms.

Through a single blood sample, he could now evaluate the relative proportions and activity of T helper cells – Th1, Th2, and Tregs – alongside cytotoxic T cells, B cells, and NK cells. These were the cells that had once fallen out of harmony, contributing to his chronic inflammation, mast cell activation, and persistent vulnerability to infections. Now, they offered a window – not just into dysfunction, but into progress.

The results told a story he already suspected. His profile still carried traces of elevated IL-4 and IL-10 activity – the hallmarks of Th2 skewing. These cytokines had long driven his hyper-reactivity, his allergic presentations, and his impaired antiviral responses. They explained his historical vulnerability to chronic infections, low NK cell activity, and autoimmunity. But now, seeing these values on paper was not about pathology. It was about clarity. About refining his recovery strategy and matching interventions to his biology, not just his symptoms.

The Th1/Th2 ratio – once an abstract concept – now carried deep personal significance. James had learned that healing didn't mean driving the immune system to an opposite extreme. It meant finding the dance between tolerance, defence, and repair. Through CD4+ T cell subtyping, he could see this ratio unfold. When it skewed low, his immune system leaned on antibodies and mast cells. When it rose gently, supported by appropriate Treg activity, he noticed fewer flares, smoother digestion, and clearer thinking. His body responded when needed – and calmed thereafter. This was intelligence. This was rhythm.

His NK cells were once almost silent. Suppressed for years by chronic stress and Th2-driven suppression, they now showed signs of recovery – an immune voice slowly returning. Interventions like DHEA, melatonin, low-dose naltrexone, and thymosin alpha-1 had supported this shift, and now, lymphocyte typing allowed him to see that change reflected in measurable terms.

His regulatory T cells – CD4+CD25+ – told a parallel story. When they were low, his world felt reactive, brittle. When they returned to balance, his tolerance did too – to food, to people, to noise, to life. He had come to understand that Tregs were not simply immune suppressors – they were the embodiment of discernment. The peacekeepers. The space between reaction and regulation. And lymphocyte typing gave him a way to see that peace, now restored, reflected in data.

As his health journey evolved, James saw how lymphocyte typing could be used not only reactively, but proactively. In the face of chronic infection, it could illuminate whether the immune system was overactive or under-responsive. In autoimmune conditions, it could provide insight into the regulatory balance needed to reduce tissue damage. And as a long-term monitoring tool, it allowed him and his clinician to revisit his profile every 9 to 12 months – checking not just for relapse, but for confirmation that *what he was doing was still working.*

Perhaps most powerful was the preventive insight. Subtle shifts in T cell ratios or NK cell numbers could alert him to impending imbalance long before he experienced symptoms. For James, this was the first time medicine felt truly personal, truly responsive – not something applied to the body, but something guided by it.

Lymphocyte typing was not just data – it was a dialogue. It told him when to support, when to pause, and most importantly, when to trust. Immune recovery wasn't a fixed destination. It was a rhythm – an ongoing conversation between environment, emotion, nutrients, pathogens, memory, and presence.

With each new set of values, James learned more about how his body danced with its environment. He no longer feared relapse. He had tools. He had insight. And now, at last, he had a map.

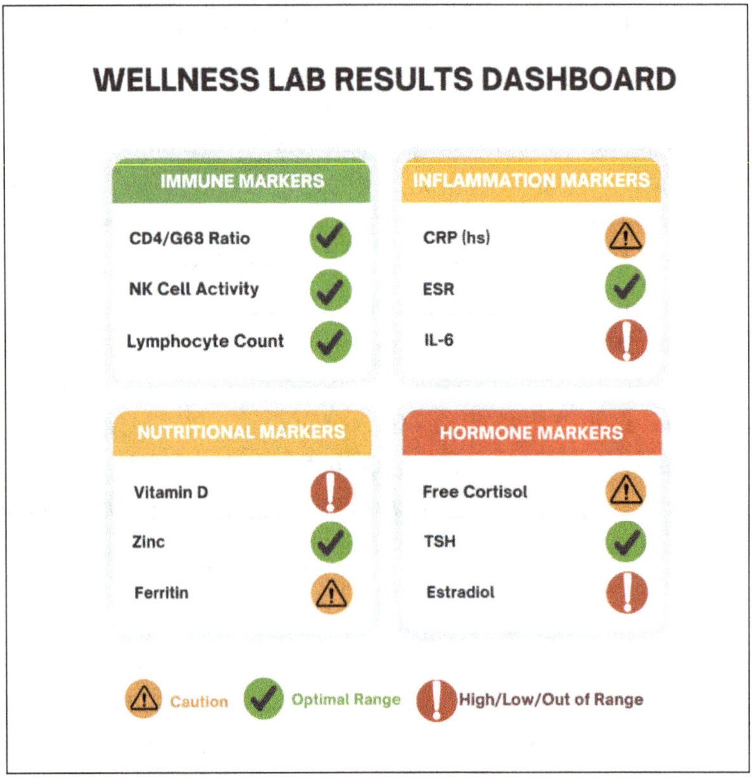
WELLNESS LAB RESULTS DASHBOARD
IMMUNE MARKERS
CD4/G68 Ratio
NK Cell Activity
Lymphocyte Count
INFLAMMATION MARKERS
CRP (hs)
ESR
IL-6
NUTRITIONAL MARKERS
Vitamin D
Zinc
Ferritin
HORMONE MARKERS
Free Cortisol
TSH
Estradiol
Caution
Optimal Range
High/Low/Out of Range

Chapter 35

Listening Beneath the Noise: Lab Testing as a Mirror to the Immune Terrain

James had come to a new place in his journey: a terrain not of diagnosis or crisis, but of clarity. He had learned that symptom was signal, but now he was learning that data, too, could be language. Labs were no longer verdicts. They were conversations. Invitations to understand which physiological systems had lost their rhythm, and why the immune system had, in its confusion, leaned into Th2 dominance as a form of misguided protection.

James began to map the body through testing – not to find what was broken, but to reveal what was burdened, blocked, or begging to recalibrate.

In the previous chapter the benefits of lymphocyte blood testing were explained. Listed below are further tests that can be undertaken for the immune system.

1. Th1/Th2 Cytokine Ratio Panel

Physiological system involved: Immune regulation via T helper cell polarisation.

Why it matters: When IL-4, IL-5, and IL-13 rise in proportion to IFNγ and IL-2, it reveals a tilt towards Th2 dominance. This means James' immune system is favouring antibody and mast cell-driven inflammation (eg allergies, asthma, eczema) over cellular immunity (eg viral clearance, tumour surveillance). The data here doesn't just show inflammation – it shows misallocation of immune resources.

2. Lymphocyte Subtyping & NK Cell Activity
Physiological system involved: Innate and adaptive immunity.
Why it matters: Th2 dominance is often accompanied by suppression of NK and cytotoxic T cells. These cells govern tumour surveillance, viral defence, and apoptosis of infected/damaged cells. James' low NK cell count told the story of an immune system that was defensive at the borders (IgE, mucosa) but blind at the core.

3. Total and Allergen-Specific IgE
Physiological system involved: B cell and mast cell-mediated humoral immunity.
Why it matters: Elevated IgE isn't just a marker of allergy – it's a confirmation of Th2 command. IgE primes mast cells for hypersensitivity, histamine surges, and barrier dysfunction (skin, lungs, gut). James' IgE levels made visible the overactive antibody arm of his immune memory.

4. Histamine and DAO Levels
Physiological system involved: Gut-immune interface and detoxification.
Why it matters: High histamine + low DAO implies poor degradation capacity, usually due to gut dysbiosis, low zinc, or inflammation. James' overactive Th2 signalling fed mast cell degranulation, while his DAO insufficiency meant he couldn't clear the result – leaving his gut and brain in a fog of irritability and inflammation.

5. Methylation Panel (SAMe, SAH, Homocysteine, B12, Folate)
Physiological system involved: Detoxification, neurotransmitter synthesis, hormone metabolism, and DNA regulation.
Why it matters: Th2-driven oxidative stress depletes methyl donors.

Low SAMe and elevated homocysteine impair oestrogen detox, glutathione production, and neurotransmitter synthesis (serotonin, dopamine). This creates a feedback loop of inflammation, mood instability, and immune misdirection.

6. Genetic Polymorphism Testing: COMT, CYP1B1, SULT1A1

Physiological system involved: Hormonal metabolism and detoxification.

Why it matters: A slow COMT or active CYP1B1 increases reactive oestrogen metabolites (eg 4-OH oestradiol), which promote oxidative stress and drive Th2 inflammation. Low SULT activity impairs sulphation, blocking clearance of histamine, hormones, and xenobiotics. For James, this helped explain the hormonal roots of his immune reactivity.

7. DUTCH Hormone Metabolite Testing (or Serum Oestradiol/ Oestrone/Metabolites)

Physiological system involved: Endocrine-immune cross-talk.

Why it matters: Elevated oestrogen metabolites stimulate mast cell activity and Th2 cytokine expression. In men and women, oestrogen dominance is a recognised driver of Th2 skew. For James, this test exposed the hormonal foundation of his Th2 reactivity, fatigue, and thyroid resistance.

8. Thyroid Panel with rT3 and Antibodies (TSH, fT3, fT4, TPO/Tg)

Physiological system involved: HPT axis and mitochondrial responsiveness.

Why it matters: Low free T3 with high rT3 reflects poor thyroid hormone utilisation – often from inflammation, low D1/D2 deiodinase activity, or elevated TBG. Elevated antibodies suggest autoimmune thyroiditis, a Th2/Th17-associated condition. Thyroid dysfunction exacerbates fatigue, poor repair, and further suppresses Th1 immunity.

9. Stool Testing: Microbiome Diversity, Beta-Glucuronidase, Histamine-Producing Species

Physiological system involved: Gut-associated lymphoid tissue (GALT), detox, and neuroimmune signalling.

Why it matters: Gut dysbiosis can promote Th2 immune responses, while high beta-glucuronidase recycles oestrogen, worsening hormonal Th2 drivers. James' elevated histamine-producing bacteria (eg Morganella, Klebsiella) kept his mast cells reactive. This test helped him restore immune tolerance through microbial recalibration.

10. Heavy Metals: Urine Provocation or Serum Lead/Mercury/Arsenic

Physiological system involved: Neuroendocrine-immune axis. *Why it matters:* Heavy metals suppress Th1 and NK activity, promoting Th2 dominance. Mercury inhibits dendritic function, impairing immune tolerance. Lead and arsenic disrupt methylation and neurotransmitters. These toxins acted as invisible architects of James' immune confusion.

11. Chronic Viral Panel: EBV, HHV-6, CMV

Physiological system involved: Latent viral burden and immune surveillance.

Why it matters: Chronic viruses like EBV produce IL-10 mimics, suppress Th1, and promote immune escape via Th2 skewing. They also deplete NK cells over time. For James, positive viral titres explained the depth of his fatigue and the persistence of his immune misdirection.

12. Vitamin D, Zinc, Selenium
Physiological system involved: Immune literacy and antioxidant buffering.
Why it matters: Vitamin D enhances Treg and Th1 balance while inhibiting IL-4/IL-13. Zinc supports DAO, Th1 skew, and epithelial integrity. Selenium is vital for thyroid and glutathione. These cofactors weren't just nutrients – they were immune teachers.

Each of these helped James understand why his immune system had chosen Th2. They translated the body's coping strategies into language he could hear, interpret, and finally, recalibrate.

In testing, not only did James find diagnosis, but also direction. And in each biomarker, he found a breadcrumb back to coherence.

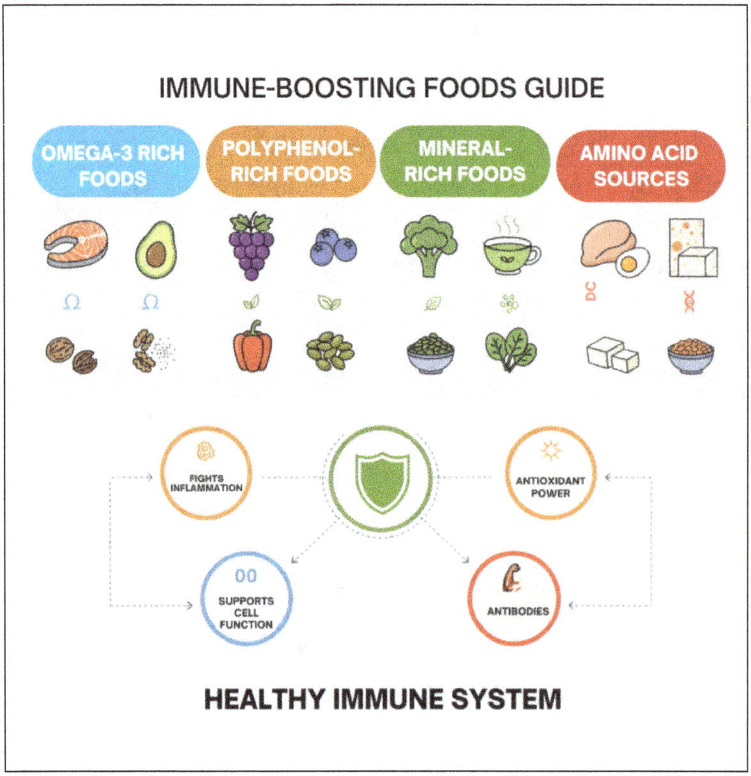
IMMUNE-BOOSTING FOODS GUIDE
OMEGA-3 RICH FOODS
POLYPHENOL-RICH FOODS
MINERAL-RICH FOODS
AMINO ACID SOURCES
FIGHTS INFLAMMATION
ANTIOXIDANT POWER
SUPPORTS CELL FUNCTION
ANTIBODIES
HEALTHY IMMUNE SYSTEM

Chapter 36

Nutritional Biochemical Considerations: The Immunological Impact of Dietary Molecules

Food represents our most continuous and concentrated interaction with the external environment. Each day, we ingest kilograms of exogenous material that our bodies must sense, detoxify, or assimilate. Historically, humans evolved to oscillate naturally between extended fasting periods and episodic feasting – a metabolic rhythm that offered survival advantages by promoting immune recalibration and metabolic flexibility.

Modern agricultural practices, however, have flattened these beneficial cycles. Year-round food abundance, aggressive crop breeding, synthetic fertilisers, and global transport have increased caloric availability but also raised exposure to inflammatory agents such as soil-depleting monocultures, organophosphate residues, ultra-processed additives, and biologically disruptive UVB radiation due to ozone-layer thinning. Unsurprisingly, the epidemiological incidence of autoimmune and metabolic disorders has expanded in parallel.

Drawing inspiration from Dr David Hawkins' insights in *Power vs. Force*, we understand nutrition as an empowering tool that invites a crucial question: Which dietary constituents most significantly burden an already toxin-saturated immune system, and how can we strategically modulate their impact?

Phenols

Phenols, generally beneficial in moderation, can exacerbate symptoms in cases of oestrogen dominance due to their interference with hormonal pathways. Dietary phenols inhibit COMT, an enzyme crucial for oestrogen metabolism and neurotransmitter clearance. Reduced COMT activity elevates oestrogen levels and neurotransmitter concentrations, exacerbating nervous system excitation, anxiety, mood instability, and cognitive disruptions.

Consequently, phenol limitation is advisable in oestrogen-dominant conditions, particularly with neurological sensitivities or mood dysregulation.

Salicylates

Salicylates exacerbate Th2 dominance by directly stimulating mast cells and basophils to release histamine and other inflammatory mediators. This activation shifts the immune balance further towards Th2 cytokine production (IL-4, IL-5, IL-13), reinforcing allergic inflammation and immune hypersensitivity.

Lectins

Lectins accelerate Th2 dominance by binding intestinal epithelial cells, compromising gut barrier integrity, and activating mast cells. This promotes increased histamine release, heightened immune activation, and amplified inflammatory responses, further driving the Th2 cytokine cascade and allergic sensitisation.

Oxalates

Oxalates intensify Th2 dominance by disrupting gut barrier integrity, irritating epithelial tissues, and triggering mast cell activation with resultant histamine release. This amplifies inflammatory responses and Th2-driven immune hypersensitivity. Additionally, oxalates bind essential minerals (calcium, magnesium, zinc), inhibiting their absorption and bioavailability, and further compromising immune regulation and cellular function.

Histamine

Histamine directly reinforces Th2 dominance by activating histamine receptors on immune cells, increasing production of IL-4 and IL-13, and promoting mast cell and basophil activation. This accentuates allergic inflammation, immune hypersensitivity, and perpetuates Th2-driven immune responses.

Matrix of Immunologically Reactive Food Molecules

The detailed matrix provided in Table 1 categorises foods by their relative content of oxalates, lectins, salicylates, histamines, and phenols, indicating high, medium, or low reactivity. It offers practical guidance for dietary modulation, emphasising strategies such as soaking, sprouting, fermenting, and high-moisture-heat cooking to reduce irritant profiles.

Table 1: Matrix of Immunologically Reactive Food Molecules (oxalate, lectin, salicylate, histamine, phenolic amines)

NUTS AND SEEDS

NUTS	Oxalate	Lectin	Salicylate	Histamine	Phenol
Almonds	High Reduce by soaking, blanching, boiling, or sprouting	High Reduce by soaking or sprouting	High	Low Small portions	High
Almond flour	High	Low	High	Low	
Almond meal		High	High	Low	
Almond butter	High	High if made from whole almonds. Low if made from soaked almonds.	High	Low	
Brazil Nuts	High Reduce by soaking, blanching, boiling or sprouting	Low	Medium	Low	
Cashews	High Reduce by soaking, blanching, boiling or sprouting	High Reduce by soaking or sprouting	High (bound sali-cylates)	High	High
Hazelnuts	Low	High Reduce by soaking or sprouting	Low	Low	High
Macadamia nuts	Low	Low	High	Low	
Pecan nuts	Low	Low	Low	Low	High
Pistachio	Medium	Low	Low	Low	High
Walnuts	Low	Low	Medium	High	High

SEEDS	Oxalate	Lectin	Salicylate	Histamine	Phenol
Chia seeds	Medium	High	Medium		
Sesame seeds	High	High	Low		
Flaxseeds	Low	Low	Medium		High
Sunflower seeds	Low	High Reduce by soaking or sprouting	Low		
Poppy seeds	Low	Low	Negligible		
Pumpkin seeds	Low	High Reduce by soaking or sprouting	High		High
Hemp seeds	Low	Low	Unknown		
Pine nuts	High	High Reduce by soaking or sprouting	High		
OTHER					
Coconut	Low	Low			

FRUITS

FRUITS	Oxalate	Lectin	Salicylate	Histamine	Phenol	Amine
Apple (canned)	Low	Low	Low	High	High	
Apple (Custard)	Low	Low	Low	Low	High	
Apple (Golden Delicious)	Low	Low	Medium	Low	High	
Apple (Red Delicious)	Low	Low	Medium	Low	High	
Apple (Pink Lady, peeled)	Low	Low	Medium	Low	High	
Apple (Pink Lady, unpeeled)	Low	Low	High	Low	High	
Avocado, unspecified variety	Low	Low	High	High		High
Apple (Granny Smith)	Low	Low	High	Low	High	
Apricot	Low	Low	High	Low	High	
Avocado (fresh), Haas	Low	Low	High	Low		
Banana	Low	Low	High	High		High
Blackberry	High	Low	High	Low	High	
Blueberry	Low	Low	High	Low	High	
Boysenberries	Low	Low	High	Low	High	
Cantaloupe (Rock melon)	Low	High	High	Low		
Cherries (canned, sweet)	Low	Low	High	Low	High	
Cherries (sour, canned, morello)	Low	Low	Medium	Low	High	
Cherries (sweet)	Low	Low	High	Low	High	
Cranberry (sauce & canned)	Low	Low	High	Low	High	
Currants	Low	Low	High	Low	High	

FRUITS	Oxalate	Lectin	Salicylate	Histamine	Phenol	Amine
Dates	High	Low	High	Low	High	
Fig (dried)	Medium	Low	High	Low	High	
Gojiberry		High		Low		
Gooseberry	Low	Low	High	Low		
Grapefruit	Low	Low	High	Low		High
Grapefruit juice	Low	Low	Low	Low		High
Grapes (fresh)	Low	Low	High	Low	High	
Grapes (green)	Low	Low	Low	Low		
Grapes, Thompson Seedless	Low	Low	High	Low		
Grapes (red)	Low	Low	High	Low	High	
Guava (canned)	Low	Low	High	Low	High	
Guava (fresh)	Low	Low	High	High	High	
Kiwi fruit	Medium	Low	High	Low		
Lemon (fresh)	Low	Low	Medium	Low		
Loganberries	Low	Low	High	Low	High	
Loquat	Low	Low	Low	Low		
Lychee	Low	Low	Medium	Low		
Mandarin	Low	Low	Medium	Low		
Mango	Low	Low	High	Low	High	
Mulberry	Low	Low	High	Low	High	
Nectarine	Low	Low	High	Low	High	
Orange	Medium	Low	High	High	High	High
Passion fruit	Low	Low	High	Low		
Pawpaw	Low	Low	Medium	High		High
Peach (fresh/ canned) (white)	Low	Low	High	Low	High	
Pear (peeled)	Low	Low	Low	High		
Pear (with peel)	Low	Low	Low	High		

FRUITS	Oxalate	Lectin	Salicylate	Histamine	Phenol	Amine
Pear (Nashi/Asian) (Packham) (William)	Low	Low	Medium	High		
Persimmon	Low	Low	High	Low		
Pineapple (canned or fresh)	High	Low	High	High		
Pineapple juice	Low	Low	Low	High		
Plums (canned)	Low	Low	High	High	High	
Plums (fresh)	Low	Low	High	High	High	
Pomegranate	Low	Low	High	Low	High	
Prunes (canned)	Low	Low	High	Low	High	
Raisins	Low	Low	High	High	High	
Raspberry (frozen or fresh)	High	Low	High	High	High	
Red currants	Low	Low	High	Low	High	
Rhubarb	Low	Low	Low	Low		
Starfruit	High	Low	No data	Low		
Strawberry	Low	Low	High	High	High	
Sultanas	Low	Low	High	Low		
Tamarillo	Low	Low	Low	Low		
Tangelo	Low	Low	High	Low		
Watermelon	Low	Low	Medium	Low		
Youngberry	Low	Low	High	Low	High	

VEGETABLES

VEGETABLES	Oxalate	Lectin	Salicylate	Histamine	Phenol	Amine
Alfalfa	Low	Low	High	Low		
Artichoke	Low	Low	High	Low	High	
Arugula / Rocket	Low	Low	High	Low		
Asparagus (canned)	Low	Low	Medium	Low		
Asparagus (fresh)	Low	Low	High	Low		
Bamboo shoot, boiled	High	Low	Negligible	Low		
Bamboo shoot, pickled	Low	Low		Low		
Beetroot	High	Low	High (Conflicting results in resources)	Low		
Bell pepper, red	Low	High Reduce with peel, deseed and cook	High	Low		
Bok choy	Low	Low	Low	Low		
Broccoli	Low	Low	Very high	Low	High	
Brussels sprouts	Low	Low	High	Low		
Cabbage, common	Low	Low	Medium	Low		
Cabbage, fresh	Low	Low	Low	Low		
Cabbage, sauerkraut	Low	Low	High	High		High
Carrots	Low	Low	Low	Low		
Cauliflower, cooked	Low	Low	Medium (Bound SA)	Low		
Cauliflower, raw	Low	Low	High	Low	High	
Celery, raw	High	Low	Medium (Bound SA)	Low	High	

VEGETABLES	Oxalate	Lectin	Salicylate	Histamine	Phenol	Amine
Chicory (fresh greens)	Low	Low	High	Low		
Chilli, red	Low	Low	High	Low		
Chinese cabbage / Napa cabbage	Low	Low	Negligible	Low		
Choko	Low	Low	Negligible	Low		
Choy sum	Low	Low	High	Low		
Courgette	Low	High	High	Low		
Cucumber, common, peeled	Low	Low	High	Low		
Cucumber, fermented pickles	Low	High	High	Low		
Cucumber, long, peeled	Low	Low	High	Low	High	
Cucumber, unpeeled	Low	High	High	Low	High	
Eggplant	Low	High	High	High	High	High
Eggplant, peeled, cooked	Low	High	Medium	High		
Eggplant, unpeeled, cooked	Low	High	High	High		
Eggplant, raw	High	High	High	High		
Endive, baby/fresh	Low	Low	High	Low		
Fennel bulbs	Low	Low	Low	Low		
Fennel tops (green parts)	Low	Low	Medium	Low		
Green beans	Low	Low	High	Low		
Green beans, cooked	Low	Low	High (cooking reduces salicylate)	Low		
Green Beans (fresh)	Low	Low	High	Low		
Green bell pepper	Low	High	High	Low	High	
Leek	Low	Low	Medium	Low		
Lettuce, butter	Low	Low	High	Low		

VEGETABLES	Oxalate	Lectin	Salicylate	Histamine	Phenol	Amine
Lettuce, cos	Low	Low	High	Low		
Lettuce, iceberg	Low	Low	Medium	Low		
Loquat	Low	Low	Medium	Low		
Mushroom, button	Low	Low	High	Low		High
Mushroom, champignon	Low	Low	High	Low		High
Mushrooms	Low	Low	Low	Low		High
Okra (canned)	High	Low	High	Low		
Olives, black (canned)	Medium	Low	High	High		
Olives, green (canned)	Medium	Low	High	High		
Onion, Spanish	Low	Low	High	Low		
Onion, white	Low	Low	High	Low		
Parsley leaves (fresh)	High	Low	Medium	Low		
Parsnip	Medium	Low	Medium	Low		
Peas, green	Low	Low	High	Low		
Peas, sugar snap	Low	Low	Medium	Low		
Pepper, green chilli	Low	High Reduce with Peel, deseed and cook	Low	Low		
Pepper, yellow green	Low	High Reduce with Peel, deseed and cook	Low	Low	High	
Pepper, green	Low	High Reduce with Peel, deseed and cook	Low	Low	High	
Pimientos (canned)	Low	Low	Low	Low		

VEGETABLES	Oxalate	Lectin	Salicylate	Histamine	Phenol	Amine
Potato, white, peeled	High	High Reduce with Peel, deseed and cook	Negligible	Low		
Potato, white, unpeeled	High	High Reduce with Peel, deseed and cook	Medium	Low		
Pumpkin marrow	Low	High Reduce with Peel, deseed and cook	Low	Low		
Pumpkin, Japanese (Kobach)	Low	High Reduce with Peel, deseed and cook	Low	Low		
Radish	Low	Low	Very high	Low		
Rhubarb	High	Low	High	Low		
Rutabaga /Swede	High	Low	High	Low		
Shallots	Low	Low	Negligible	Low		
Spinach	High Lower if cooked	Low	High	High	High	
Spinach, frozen	Low	Low	Low	Low		
Squash, baby	Low	High Reduce with Peel, deseed and cook	Low	Low		
Squash, butternut	Low	High Reduce with Peel, deseed and cook	Low	Low		
Sweet potato, white	High	Low	High	Low		

VEGETABLES	Oxalate	Lectin	Salicylate	Histamine	Phenol	Amine
Sweet potato, yellow	High	Low	Very high	Low		
Tomato soup	Low	High	High	High	High	High
Tomato (canned), diced	Low	High	High	High	High	High
Tomato, cherry	Low	High Reduce with Peel, deseed and cook	Low	High	High	High
Tomato, common, beefsteak	Low	High Reduce with Peel, deseed and cook	Low	High	High	High
Tomato, paste / sauce / canned / sundried	Low	High Reduce with Peel, deseed and cook	Very high	High	High	High
Tomato, peeled	Low	High Reduce with Peel, deseed and cook	High	High	High	High
Tomato, Roma	Low	High Reduce with Peel, deseed and cook	Low	High	High	High
Tomato, unpeeled	Low	High	High	High	High	High
Turnip	High	Low	Low	Low		
Watercress	Low	Low	High	Low		
Yellow beans	Low	Low	Very high	Low		

VEGETABLES	Oxalate	Lectin	Salicylate	Histamine	Phenol	Amine
Zucchini / Courgette	Low	High Reduce with Peel, deseed and cook	Low	Low	High	

FATS AND OILS

FATS & OILS	Oxalate	Lectin	Salicylate	Histamine	Phenol
Ghee	Low	Low	Low	Low	
Olive oil (Balducci, Moro, Bertolli)	Low	Low	Low	Low	
Palm oil	Low	Low	Low	Low	
Almond oil	Low	Low	Medium	Low	
Walnut oil	Low	Low	High	Low	
Coconut oil	Low	Low	Very high	Low	
Macadamia nut oil	Low	Low	Low	Low	
Avocado oil	Low	Low	Low	Low	

BEVERAGES

BEVERAGES	Oxalate	Lectin	Salicylate	Histamine	Phenol	Amine
Black tea	Low in a single cup with medium brewing time. Longer steeping times would increase oxa-late content	Low	Low	Low		High
Instant coffee	Low	Low	Low	Low		High
Coffee, brewed from beans	Low	Low	Medium	Low		High
English tea	Low	Low	Low	Low		

BEVERAGES	Oxalate	Lectin	Salicylate	Histamine	Phenol	Amine
Peppermint tea	Low	Low	Low	Low		
Tea bag, Earl Grey, English Breakfast, Irish Breakfast	Low	Low	High	Low		
Chamomile tea	No reliable data		Medium	Low		
Herbal fruit tea	Low	Low	Medium	Low		
Rosehip tea	Low	Low	Medium	Low		

MISCELLANEOUS

MISCELLANEOUS	Oxalate	Lectin	Salicylate	Histamine	Phenol	Amine
Apple butter, homemade from acceptable peeled apples	Low	Low	Negligible	Low		
Saffron	Low	Low	Negligible	Low		
Vinegar, malt	Low	Low	Negligible	High		
Baking, Bi-carb Soda	Low	Low	Negligible	Low		
Baking powder	Low	Low	Negligible	Low		
Carob powder	Low	Low	Negligible	High		
Citric acid	Low	Low	Negligible	Low		
Cocoa powder	Low	Low	Negligible	High	High	High
Cream of tartar	Low	Low	Negligible	Low		
Maple syrup	Low	Low	Negligible	Low		
Pectin	Low	Low	Negligible	Low		
Psyllium husks	Low	Low	Negligible	Low		
Salt	Low	Low	Negligible	Low		
Coriander leaves, fresh	High		Low	Low		
Basil leaves, fresh	Low	Low	Medium	Low		
Lemon grass	Low	Low	Medium	Low		
Fennel powder	Low	Low	High	Low		

MISCELLANEOUS	Oxalate	Lectin	Salicylate	Histamine	Phenol	Amine
Coconut flakes	Low	Low	High	Low		
Honey	Low	Low	High	Low		
Vinegar, balsamic	Low	Low	High	High		High
Vinegar, white / wine	Low	Low	High	High		High
All spice powder	Low	Low	High	Low		
Aniseed powder	Low	Low	High	Low	High	
Basil leaves, powder	Low	Low	High	Low		
Bay leaf	Low	Low	High	Low		
Canella, powder	Low	Low	High	Low		
Cardamom, powder	Low	Low	High	Low		
Cayenne, powder	Low	Low	High	Low	High	
Celery, powder	Low	Low	High	Low	High	
Chilli flakes	Low	Low	High	Low		
Chilli powder	Low	Low	High	Low		
Chives, fresh	Low	Low	High	Low		
Cinnamon powder	Low	Low	High	Low	High	
Cumin	Low	Low	High	Low	High	
Curry powder	Low	Low	High	Low	High	
Dill, fresh	Low	Low	High	Low	High	
Dill, dried	Low	Low	High	Low	High	
Fenugreek powder	Low	Low	High	Low		
Garam masala powder	Low	Low	High	Low		
Garlic, fresh	Low	Low	High	Low		
Ginger, fresh	Low	Low	High	Low		
Mace powder	Low	Low	High	Low	High	
Mint, fresh garden	Low	Low	High	Low		
Mustard powder	Low	Low	High	Low	High	High
Nutmeg powder	Low	Low	High	Low		
Oregano powder	Low	Low	High	Low	High	
Paprika, hot powder	Low	Low	High	Low	High	
Paprika, sweet powder	Low	Low	High	Low		

MISCELLANEOUS	Oxalate	Lectin	Salicylate	Histamine	Phenol	Amine
Pepper, black powder	Low	Low	High	Low		
Pepper, white powder	Low	Low	High	Low		
Pimiento powder	Low	Low	High	Low		
Rosemary powder/ fresh	Low	Low	High	Low	High	
Sage leaves	Low	Low	High	Low	High	
Tarragon powder	Low	Low	High	Low	High	
Thyme leaves	Low	Low	High	Low	High	
Turmeric powder	Low	Low	High	Low	High	
Vanillin					High	

EGGS

EGGS	Oxalate	Lectin	Salicylate	Histamine	Phenols
Egg (white + yolk)	Low	Low	Negligible	Low	

POULTRY

POULTRY	Oxalate	Lectin	Salicylate	Histamine	Phenol	Amine
Chicken / Turkey	Oxalate not present	No evidence that grain fed produce would have lectin content in the product	No salicylate present in chicken or turkey Possibly high in canned poultry with seasonings Paste / Spreads with seasonings	Low: Natural fresh meat from chicken or turkey Frozen poultry, thawed rapidly High: any processed chicken products with yeast / monosodium glutamate Canned meat, cured, marinated, smoked or preserved products		Cured and processed poultry such as hot dogs, ham and bacon/ macon

RED MEATS

RED MEAT	Oxalate	Lectin	Salicylate	Histamine	Phenol	Amine
Beef	Oxalate not present		Luncheon meats Canned meats with flavourings Meat paste with seasonings	Low: Natural fresh meat Frozen, thawed rapidly High: any processed meat products with yeast / mono-sodium glutamate Canned meat, cured, marinated, smoked or preserved products Bone-matured or dry-aged meat, long-hung meat (especially beef) Finely chopped/ pureed/ ground meat (meatloaf, spreads, cold cuts) Histamine content rise with the degree of comminution Luncheon meats Meat paste		Cured and processed meats such as salami, hot dogs, ham and bacon

RED MEAT	Oxalate	Lectin	Salicylate	Histamine	Phenol	Amine
Pork				Canned meat, cured, marinated, smoked, dry-cured ham, bacon or preserved products		

FISH

FISH	Oxalate	Lectin	Salicylate	Histamine	Phenol	Amine
Fish	Low	Low	Negligible in fresh / frozen fish	High: tuna, mackerel, herring, sardines, anchovies, salmon Low: absolutely fresh caught fish, frozen fish (no long-time standing), quick thaw and use immediately, eg pollock cod, trout, whitefish, organic pangasius		
Shellfish			Negligible	High: mussels, lobsters, crabs, shrimps, prawns		
Fish (canned in oil/ seasoning)			High	High		

FISH	Oxalate	Lectin	Salicylate	Histamine	Phenol	Amine
Canned, marinated, salted, dried, smoked or pickled fish or seafood			High	High		High

Transforming Irritant Profiles Through Preparation

Simple culinary techniques significantly reshape a food's immunological impact.

For oxalate-rich vegetables, a brief 10-minute boil can leach up to 80 per cent of soluble oxalates like spinach or beet greens into the cooking water, whereas steaming provides a modest reduction (5–50 per cent) and baking yields minimal change.

Lectin loads – particularly in nuts, seeds, and nightshades – drop substantially when outer coatings are peeled and deseeded, with further reductions achievable via soaking, sprouting, and fermenting. High-moisture-heat methods (eg soups and stews) denature remaining lectins effectively. Salicylate concentrations, notably high in tomato products and berries, decrease significantly when fruits or vegetables are simmered with cooking liquids discarded. Histamine and biogenic amines, though less heat sensitive, can be managed by rapid freezing post-harvest and avoiding prolonged marination or slow fermentation of animal proteins.

Thus, preparation is a potent, evidence-backed strategy for reducing biochemical burdens, transforming many 'high' reactivity foods into moderate or low options without sacrificing dietary diversity or enjoyment.

Through these nutritional biochemical considerations, James found a crucial lever for fine-tuning his immune resilience – not through dietary restriction alone, but through informed, intelligent modulation of dietary immunological loads.

THE IMMUNE WELLNESS WHEEL

Chapter 37

Conclusions: The Return to Immune Rhythm

James had entered his healing journey searching for relief – from allergies that disrupted his daily life, from fatigue that robbed him of clarity, and from inflammation that kept his body in a state of readiness long after the threat had passed. What he uncovered along the way was not just a diagnosis or a protocol, but a complete re-education: an understanding of how his immune system had been shaped, how it had adapted, and how it could be gently rewired into balance.

At the centre of it all was Th2 dominance – a skewed immune profile driven by overexpression of cytokines like IL-4, IL-5, and IL13, fostering heightened humoral activity, mast cell instability, and systemic inflammation. This pattern, James discovered, did not exist in isolation. It intersected with mitochondrial dysfunction, thyroid hormone resistance, impaired detoxification, histamine overload, and nervous system dysregulation. Th2 dominance was not just an immune imbalance. It was a web of disrupted communication between the immune, endocrine, and neurological systems.

The path forward was not linear. James learned to work in layers – rebalancing cytokines through micro-immunotherapy and low-dose naltrexone, both of which subtly downregulated Th2 signalling while enhancing regulatory T cell function and restoring immune tolerance. He found support through thymosin alpha-1, a peptide that enhanced innate immunity and improved cellular communication without overstimulation, giving his natural killer cells and dendritic network the strength they had long lacked.

But immune signalling on its own was not enough. James realised that his recovery depended on improving endocrine receptivity and mitochondrial vitality. This meant optimising thyroid hormones – not simply with medication, but with cofactors like selenium, zinc, and iodine, which allowed T3 to interact with its mitochondrial receptors.

Mitochondrial function itself was enhanced through targeted nutrients like CoQ10, PQQ, and resveratrol – tools that restored redox balance, improved ATP production, and softened the cellular rigidity that kept his metabolism sluggish and inflamed.

Fasting became another chapter in James' regulation. Short, cyclic fasts allowed him to reduce IL-6 and TNF-α, activate autophagy, and reshape his microbiome. He came to appreciate that fasting was less about restriction and more about rhythm – a space for immune recalibration, neuroimmune coherence, and a pause in the constant input that once kept his mast cells over-alert.

Diet played its part too. Moving towards anti-inflammatory, low-histamine foods rich in omega-3s gave his gut and immune system the signals they needed to lower the baseline noise of histamine overload. It was not about purity – it was about *precision*. Knowing what supported his system and what pushed it into overdrive.

Mast cell stabilisation became the backbone of his symptomatic management. Through cromolyn sodium, ketotifen, and enzymes like Bromelain, he found relief from histamine surges that once defined his life. Unlike suppressive medications, these tools helped restore the threshold – keeping mast cells grounded, reducing IgE responses, and calming cytokine cascades that once spiralled out of control.

Alongside those, newer agents like Maraviroc and Cepharanthine brought something more: the ability to modify immune trafficking, to intercept chemokine signals, and to rebalance redox and membrane fluidity. These deeper interventions addressed the architecture of inflammation – not just the chemicals, but the communication networks between immune cells and tissues. Through them, James felt his reactivity dissolve into resilience.

He learned the value of tailoring each intervention to his unique genetic, hormonal, and detoxification profile. What worked for one stage of healing needed refinement in the next. He monitored lymphocyte subsets, cytokine ratios, and thyroid function regularly – not out of fear, but out of partnership with his body. He wasn't just treating illness anymore. He was supporting function.

The final layer – perhaps the most essential – was the nervous system. Through vagal toning, limbic retraining, and somatic practices, James slowly unwound the defensive tone that had gripped his physiology for years. It was here, in the stillness between sympathetic activation and parasympathetic rest, that he found the full integration of his work. Healing was not about fixing – it was about teaching his body how to trust again.

James didn't need perfection to feel whole. He needed a system that could *respond appropriately*, recover quickly, and communicate clearly. And now, his immune system had become just that: adaptable, responsive, and informed.

Th2 dominance had once dictated his experience of the world. Now, it had become a teacher – a lens through which he learned how deeply interconnected all systems are, and how powerfully healing unfolds when those systems are given the chance to listen, repair, and reorient.

Immune balance was not a supplement. It was a *state of coherence* – nourished by information, refined through experience, and grounded in rhythm.

James had not simply resolved a condition.

He had reclaimed his *capacity for resilience*.

Epilogue:

Living Beyond the Pattern

James no longer thought in terms of dominance or deficiency. The language of 'fixing' had faded. What remained was a quiet knowing – a felt sense that the body, once lost in chaos, had found its rhythm again.

There were still days of fluctuation. He expected them. But what once would have sparked panic, or spiralled into symptom stacking, now became *an invitation to pause*, to observe, to recalibrate with respect. His immune system was no longer his adversary. It was a messenger, a compass, a partner.

He had moved beyond the pattern.

Not because it had disappeared entirely, but because it no longer defined the architecture of his choices.

Where he once feared food, he now ate with awareness. Where he once dreaded social exposure, he now moved through it with presence. Where he once woke up bracing for the day, he now rose into it with a nervous system that had learned *to rest without shutting down* and *to respond without overreacting.*

He now understood that resilience was not an absence of inflammation or an eradication of cytokines. It was a *return to centre after the immune system had done its work*. It was the *grace between activation and resolution*.

And this grace – this space – was something James had built.

Through targeted therapies, mitochondrial repair, neuroimmune integration, and a nervous system that finally trusted safety, James had arrived at a new terrain – one where he no longer lived in pursuit of health, but in *collaboration with it.*

He no longer needed to micromanage every cytokine. He had tools, yes. He had data, yes. But more than that, he had a *relationship*. With his biology. With his symptoms. With his self.

Healing had never been about fixing what was broken.

It was about *recovering what was wise*.

And now, as he moved into life beyond the pattern, James didn't see an endpoint. He saw a rhythm. A pulsing invitation to stay attuned, to adjust when needed, and to return, always, to that quiet place where breath deepens, cells repair, and the immune system learns to listen again.

He wasn't just better.

He was *home*.

Notes on Protocols

Th2 dominance is a paradigm involving multiple physiological contributors, all of which – as discussed throughout this book – are notably influenced by environmental and psychological factors. The answers to this anomaly are not found solely in diagnostic outcomes; rather, understanding Th2 dominance requires consideration of both objective parameters (such as lab results) and the individual's subjective experience. Only through a comprehensive evaluation of the individual can we hope to establish a blueprint in which carefully implemented strategies may catalyse a healing trajectory.

Below, I have provided multiple intervention aid protocols relevant to the numerous contributors to Th2 dominance. May they inspire motivation – offering adaptive insight and helping to develop a unique remedy for each person struggling with this condition.

NK Cell Insufficiency

Immunological surveillance is one of the first systems to exhibit insufficiency in Th2 dominance, often due to reduced natural killer cell activity. The following protocols offer three options to support NK cell function, depending on iron status and oestrogen metabolism. In my observation, the Specific Carbohydrate Diet yields the most benefit when implementing any strategy aimed at improving NK cell density.

https://autonomicwellness.com/specific-carbohydrate-diet/

<u>Protocol One:</u> For those without any history of iron anaemia or oestrogen dominance.

Rx.

Duration: 4–8 weeks – pending lab review.

- Imunovir

Week one: 500 mg three times per day (TID) Mon–Fri
Week two: 500 mg once per day (QD) Mon–Fri
Week three and four: None
Repeat twice then break for 4 weeks before resuming if needed.

- Cepharanthine

Week one and two: none
Week three and four: 5 mg twice per day (BID) Mon–Fri

Micro-Immunotherapy

- MISEN 1 capsule QD on waking
- EID (-N) 1 capsule 3 hours after waking

BHRT.

- Biomatrix DHEA 2 drops TID
- Melatonin 300 mcg–3 mg QD

Nx.

- Neogenesis IP6 5 gr TID
- Coyne Health AHCC 1 gr TID
- Unmatched Immuno IgG 1 scoop QD

Peptides

- TA1 1.5 mg every three days (Q3D)

<u>Protocol Two:</u> For those without a history of anaemia but have experienced oestrogen dominance.

Rx.

Duration: 4–8 weeks – pending lab review.

- Imunovir

Week one: 500 mg TID Mon–Fri
Week two: 500 mg QD Mon–Fri
Week three and four: None
Repeat twice then break for 4 weeks before resuming if needed.

- Cepharanthine

Week one and two: none
Week three and four: 5 mg BID Mon–Fri

Micro-Immunotherapy

- MISEN 1 capsule QD on waking
- EID (-N) 1 capsule 3 hours after waking

BHRT.

- Melatonin 300 mcg–3 mg QD

Nx.

- Neogenesis IP6 5 gr TID
- Coyne Health AHCC 1 gr TID
- Unmatched Immuno IgG 1 scoop QD
- Calcium D-glucarate 1500 mg BID

Peptides

- TA1 1.5 mg Q3D

<u>Protocol Three:</u> for those who have experienced iron anaemia.

Rx.

Duration: 4–8 weeks – pending lab review.

- Imunovir

Week one: 500 mg TID Mon–Fri
Week two: 500 mg QD Mon–Fri
Week three and four: None
Repeat twice then break for 4 weeks before resuming if needed.

Micro-Immunotherapy

- MISEN 1 capsule QD on waking
- EID (-N) 1 capsule 3 hours after waking

BHRT.

- Melatonin 300 mcg–3 mg QD

Nx.

- 4Life Transfer factors plus 2–6 TID (slowly titrate tolerance)
- Coyne Health AHCC 1 gr TID
- Unmatched Immuno IgG 1 scoop QD
- Calcium D-glucarate 1500 mg BID

Peptides

- TA1 1.5 mg Q3D

Heightened Histamine

Elevated histamine levels are synonymous with Th2 dominance, contributing not only to an increased likelihood of rhinitis but also to disrupted sleep quality and circadian rhythm. Histamine is released by the immune system but is also present in various foods. Therefore, the following protocol addresses histamine metabolism through both immunological modulation and digestive support.

From a nutritional perspective, a low-histamine Palaeolithic plan can go a long way in restoring histamine tolerance. In more severe cases, this may be further enhanced by coupling a low-histamine intake with supported intermittent fasting.

https://autonomicwellness.com/low-histamine-palaeolithic-plan/

Protocol Histamine

Period of application: 8 weeks

Rx.

(Note: not for those with hypochlorhydria aka low stomach acid)

- Ketotifen 1 mg QD-TID (note: not for those with low free cortisol or slow CYP3A4 liver enzyme activity, in these cases ketotifen can induce depression)
- Cromolyn Sodium 100 mg QD-TID 30 min before meals

Nx.

- Seeking Health HistaminX (note: not for those with sulphur sensitivity) 1–2 capsules BID – away from food
- Seeking health HistamineDigest 1–2 daily 30 min before meals
- Seeking health Probiota HistaminX 1–5 capsules daily, start slow and titrate

Peptides

- LVLUP Larazotide 500 mcg BID

Spleen Insufficiency

During Th2 dominance, the spleen can become congested – often as a result of methylation imbalances caused by elevated histamine levels. This congestion may impair antioxidant activity and contribute to systemic oxidative stress. The following protocol is designed to support spleen function, thereby enhancing both antioxidant status and mitochondrial electron transport chain activity.

Protocol: spleen reset

Micro-immunotherapy

- MISEN 1 capsule on waking
- MIREG 1 capsule 3 hours after waking

Nx.

- Researched Nutritionals transfer factor multi-immune 1–4 capsules TID (start slow and titrate)
- Xymogen Adrenal Manager 1–2 capsules BID: morning and midday
- Objective Nutrients ThiAssist 1 scoop BID: morning and midday
- Objective Nutrients ThiActive B 2 capsules BID: morning and midday
- Objective Nutrients ThiaMax 1 capsule BID: morning and midday
- QuickSilver Scientific molecular hydrogen 1–4 daily

Peptides

Epitalon 1 gr QD

Adrenal Health

From viral contributors to heavy metals, various factors not only influence Th2 dominance but also affect adrenal health. The following protocol is designed to support individuals with low cortisol output. Notably, the most effective nutritional plan I have observed for supporting diminished adrenal tone is the carnivore diet – although its application requires careful monitoring of gastrointestinal symptoms, particularly changes in bowel transit time.

https://autonomicwellness.com/carnivore-diet/

There are varying degrees of impaired cortisol function, each best assessed through a DUTCH Plus test – or, at minimum, a salivary cortisol panel. The following protocols are structured to address cortisol insufficiency ranging from severe to mild.

Protocol One: severe cortisol insufficiency.
Period of application: 3–6 weeks *rechecking labs after every 3 weeks.

Rx.

- Imunovir

Week one: 500 mg TID Mon–Fri
Week two: 500 mg QD Mon–Fri
Week three and four: None
Repeat twice then break for 4 weeks before resuming if needed.

BHRT.

- Regenerative Nutrition Adrenal Max +* (determined on the following saliva cortisol parameters):

- Morning levels (5–7 am)
    - 10-13nM = 1 capsule
    - 5-9nM = 2 capsules
    - <5nM = 3–4 capsules
- Late morning levels (10–11 am)
    - <4nM = 1–2 capsules
- Afternoon levels (2–3 pm)
    - <3nM = 1 capsule

*avoid licorice root with this product

Nx.

- Xymogen Adrenal Manager 1–2 capsules BID without food
- Xymogen MinRex 3–4 capsules QD with food

Peptides

- Epitalon 10 mg QD for 10 days then 1 mg QD
- Intranasal SEMAX 200mcg BID

Protocol Two: moderate cortisol insufficiency.
Period of application: 3–6 weeks *rechecking labs after every 3 weeks

Rx.

- Imunovir

Week one: 500 mg TID Mon–Fri
Week two: 500 mg QD Mon–Fri
Week three and four: None
Repeat twice then break for 4 weeks before resuming if needed.

BHRT.

- Biomatrix Pregnenolone
  - 5–7 am: 5–20 drops (depending on severity of symptoms of fatigue)
  - 10–11 am: 5–10 drops
  - 2–3 pm: 2–5 drops

- Biomatrix DHEA (only use if morning symptoms alleviate)
  - 5–7 am: 1–2 drops
  - 10–11 am: 1–2 drops
  - 2–3 pm: 1–2 drops

Nx.

- Biomatrix Support Adrenals 1–2 capsules BID with food
- Biomatrix Support Minerals 3–4 capsules QD with food
- Biomatrix licorice root: gauge stimulation, too much will negatively impact anxiety and sleep
  - 5–7 am: 5–15 drops
  - 10–11 am: 5–10 drops
  - 2–3 pm: 2–5 drops (if needed)

Peptides

- Intranasal Selank 500 mcg BID

Protocol Three: minor cortisol insufficiency.

Nx.

- Biomatrix Support Adrenals 1–2 capsules BID with food
- Biomatrix Support Minerals 3–4 capsules QD with food
- Biomatrix licorice root: gauge stimulation, too much will negatively impact anxiety and sleep
- 5–7 am: 5–15 drops
- 10–11 am: 5–10 drops
- 2–3 pm: 2–5 drops (if needed)

Leaky Gut

Gastritis as well as leaky gut are intimately associated with heightened levels of histamine. The following protocol is designed to address gastric inflammation and leaky gut syndrome.

Protocol: leaky gut

Rx.

- Ketotifen 1 mg TID (note above mention on case tolerance)

BHRT.

- DHEA (if there is no history of oestrogen dominance) 20–100 mg TID
- Melatonin 20–200 mg QD (if there are no issues with free cortisol)

Nx.

- Microbiome labs megamucosa 1 scoop 1–4 times daily. Start slow and titrate
- Devrom 1 tablet four times per day (QID) away from food
- Sodium Butyrate 600–1200 mg 2–4 times daily with food
- Designs for Health GastroMend 3–6 capsules QD

Peptides

- LVLUP Larazotide 500 mcg BID

Candida

Elevated histamine levels, when combined with compromised mitochondrial efficiency, can significantly increase the body's susceptibility to fungal overgrowth – particularly *Candida*. As *Candida* proliferates, biofilm accumulation follows, further disrupting the commensal bacterial profile and perpetuating a cycle of chronic candidiasis. The following protocol offers a simple yet effective approach to addressing *Candida* and related fungal overgrowth.

Rx.

- Nystatin 500 mg BID
- Ketotifen 1 mg QD

Nx.

- Biocidin LSF 1–3 pumps BID (30 min before food)
- Biocidin GI detox 2 capsules 90 min away from food or other supplements/Rx
- Saccharomyces Boulardii 250–500 mg QD taken before sleep

- Magnesium Citrate 75–300 mg TID with meals

Peptides

- LVLUP Ultimate GI repair without BPC157 2 capsules QD

SIBO (Small Intestinal Bacterial Overgrowth)

Our microbiomes exhibit 150X more DNA expressions than our human cells; therefore, when there is a disturbance in our immunological paradigm, outcomes such as SIBO are commonly experienced. Below you will find my favourite protocol to address SIBO. Additionally, I have found the FODMAP diet to work best for SIBO. ***https://autonomicwellness.com/fodmap-diet/***

Period of application: 8 weeks

Rx.

- Rifaximin 200 mg TID Q3D

Nx.

- BodyBio PC 1.2 gr BID
- BodyBio Sodium Butyrate 600–1200 mg BID
- Biocidin GI detox 2 capsules BID away from food
- ION gut minerals 1–3 tsp BID

Peptides

- LVLUP Ultimate GI repair 2 capsules BID

Oestrogen Dominance

Hormonal imbalances often stem from disruptions in the body's ability to metabolise and detoxify hormones effectively. In the context of Th2 dominance, resulting outcomes such as oestrogen dominance can further exacerbate the paradigm – particularly by interfering with thyroid receptor sensitivity and hormone activity.

Below, you will find my preferred protocols for addressing oestrogen dominance in both men and women. Notably, a low-carbohydrate Palaeolithic diet has consistently proven beneficial as a foundational dietary strategy within this context.

https://autonomicwellness.com/low-carbohydrate-palaeolithic/

Protocol: Men oestrogen dominance

Period of application: 8 weeks

Rx.

- Glucophage 500XR QD
- Glucobay 50 mg BID (before meals)

Nx.

- Biomatrix Support Glucose 1–2 capsules BID (30 min before meals)
- Calcium D-glucarate 500–3000 mg BID
- DIM 100–400 mg BID
- BodyBio PC 1.2 gr BID
- BodyBio Sodium Butyrate 1.2 gr BID
- LVLUP Liver Complex 3 capsules QD

Peptides

- Ipamorelin 100 mcg BID
- Tesamorelin 250 mcg BID (precaution: increases of IGF-1 are associated with oestrogen dominance)

Protocol: Women oestrogen dominance

Rx.

- Rapamune 1 mg every seven days (Q7D) (not for those immune compromised or prone to migraine)
- Glucophage 500XR QD
- Glucobay 50 mg BID (before meals)

Nx.

- Vitanic Fem-rebalance 2 capsules QD days 10–2 of cycle
- Thorne Homocysteine supreme QD days 14–28 of cycle
- Thorne Stress B complex 1–3 capsules QD days 1–13 of cycle
- Calcium D-glucarate 500–1500 mg BID days 14–28 of cycle
- DIM 100–200 mg BID days 14–28 of cycle (caution in hypothyroid states)
- Integrative Therapeutics Blue Heron 6 capsules QD

Peptides

- Selank 500 mcg BID

Macrophage Activation Syndrome

In extreme cases of chronic immune compromise, Th2 dominance not only becomes a pronounced reality, but Macrophage Activation Syndrome (MAS) may also emerge as a critical concern. The following protocol is intended to support individuals experiencing autoimmune symptoms associated with Th2-dominant CD4+ T cell disruption.

From a nutritional standpoint, the Specific Carbohydrate Diet (SCD) stands out – by a country mile – as the most effective strategy in this context.

https://autonomicwellness.com/specific-carbohydrate-diet/

Note: *Lymphocyte typing and cytokine mapping should be conducted at 8-week intervals to monitor immunological shifts and guide ongoing protocol refinement.*

Period of application: 8–16 weeks

Protocol: complex MAS and Th2 cases

Rx.

- Maraviroc 150–300 mg BID Q3D (eosinophilia is a probable concern with this Rx. so avoid intake if historical hepatic or allergic sensitivity is the case).
- Ketotifen 1 mg TID
- Imunovir

- Week one: 500 mg TID Mon–Fri
- Week two: 500 mg QD Mon–Fri
- Week three and four: None

- Repeat twice then break for 4 weeks before resuming if needed.

- Cepharanthine

- Week one and two: none
- Week three and four: 5 mg BID Mon–Fri

Nx.

- Neogenesis IP 6 5 gr BID
- AHCC 1 gr TID
- Researched nutritionals transfer factor immune 2–6 capsules BID
- Sodium Butyrate 1.2 gr BID
- Devrom 1 tab QID away from food
- ION Gut minerals 1 tsp TID
- Thorne Stress B complex 1–6 capsules QD
- Neogenesis LifeSpan 2 capsules QD
- Researched nutritionals ATP fuel 5 capsules BID

Peptides

- HEP1 2 mg QID
- TA1 1 mg QD

Oral Health

Dr Dominik Nischwitz's Oral Health protocol:

- **Clean toothpaste** – Use a toxin-free toothpaste like More Than a Toothpaste from SUPZ Nutrition. Brush either upon waking or at least 30 minutes after food.
- **Copper tongue scraper** – Every morning, scrape your tongue for 10–15 seconds to remove bacteria and freshen breath.
- **Coconut oil pulling** – Use instead of chemical mouthwash. Take 5–15 minutes while doing other tasks, then spit it into the bin, not the sink. It is antimicrobial, antiviral, and soothing to gums and teeth.
- **Support from within** – Take Bone & Teeth from SUPZ Nutrition: 3 capsules every evening to nourish teeth and bone from the inside out.

Abbreviations

ATP:	Adenosine triphosphate
BCR:	B cell receptor
BID:	Twice per day
CDR:	Cell Danger Response
COMT:	Catechol-O-methyltransferase
CRH:	Corticotropin-releasing hormone
CTL:	Cytotoxic T Lymphocytes
DAMP:	Damage-Associated Molecular Patterns
DAO:	Diamine oxidase
DHEA:	Dehydroepiandrosterone
DIM:	Diindolylmethane
EBV:	Epstein-Barr virus
FDA:	Food and Drug Administration
FDOJ:	Fatty Degenerative Osteonecrosis of the Jaw
GALT:	Gut-associated lymphoid tissue
GLP-1:	Glucagon-Like Peptide-1
HPA:	Hypothalamic-pituitary-adrenal
HPT:	Hypothalamic-pituitary-thyroid
IgE:	Immunoglobulin E
IFN-γ:	Interferon-gamma
iTREGS:	Induced regulatory T cells
IL:	Interleukin
LDN:	Low-dose naltrexone
MAS:	Macrophage Activation Syndrome
MHC II:	Major Histocompatibility Complex class II
NICO:	Neuralgia-Inducing Cavitational Osteonecrosis

NK:	Natural Killer
NOS:	Nitric oxide synthase
POP:	Persistent organic pollutants
QD:	Once per day
QID:	Four times per day
ROS:	Reactive oxygen species
SAMe:	S-adenosylmethionine
SCD:	Specific Carbohydrate Diet
SHBG:	Sex hormone-binding globulin
SIBO:	Small intestinal bacterial overgrowth
SOD:	Superoxide dismutase
SULT:	Sulfotransferase enzymes
T3:	Triiodothyronine
T4:	Thyroxine
Tα1:	Thymosin alpha-1
TBG:	Thyroid-binding globulin
TID:	Three times per day
TLR:	Toll-like receptors
TRH:	Thyrotropin-releasing hormone
TSH:	Thyroid-stimulating hormone

References

Abel, S., Back, D. J., and Vourvahis, M. (2009) Maraviroc: Pharmacokinetics and drug interactions. *Antiviral Therapy*, 14(5), 607–618. https://doi.org/10.1177/135965350901400514

Abel, S., Van Der Ryst, E., Rosario, M.C., Ridgway, C.E., Medhurst, C.G., Taylor-Worth, R.J. and Muirhead, G.J. (2008) Assessment of the pharmacokinetics, safety and tolerability of maraviroc, a novel CCR5 antagonist, in healthy volunteers. *British journal of clinical pharmacology*, *65*, pp.5-18. https://doi.org/10.1111/j.1365-2125.2008.03130.x

Aggarwal, B.B. and Harikumar, K.B. (2009) Potential therapeutic effects of curcumin, the anti-inflammatory agent, against neurodegenerative, cardiovascular, pulmonary, metabolic, autoimmune and neoplastic diseases. *The International Journal of Biochemistry & Cell Biology*, 41(1), pp.40-59. https://doi.org/10.1016/j.biocel.2008.06.010

Akdis, C.A., Arkwright, P.D., Brüggen, M.C., Busse, W., Gadina, M., Guttman-Yassky, E., Kabashima, K., Mitamura, Y., Vian, L., Wu, J. and Palomares, O. (2020) Type 2 immunity in the skin and lungs. *Allergy*, *75*(7), pp.1582-1605. https://doi.org/10.1111/all.14318

Alp, N.J. and Channon, K.M. (2004) Regulation of endothelial nitric oxide synthase by tetrahydrobiopterin in vascular disease. *Arteriosclerosis, thrombosis, and vascular biology*, *24*(3), pp.413-420. https://doi.org/10.1161/01.ATV.0000110785.96039.f6

Bailly, C. (2019) Cepharanthine: An update of its mode of action, pharmacological properties and medical applications. *Phytomedicine*, 62, p.152956. https://doi.org/10.1016/j.phymed.2019.152956

Bottiglieri, T. (2002) S-Adenosyl-L-methionine (SAMe): from the bench to the bedside—molecular basis of a pleiotrophic molecule. *The American journal of clinical nutrition*, 76(5), pp.1151S-1157S. https://doi.org/10.1093/ajcn/76.5.1151S

Cajochen, C., Kräuchi, K. and Wirz-Justice, A. (2003) Role of melatonin in the regulation of human circadian rhythms and sleep. *Journal of neuroendocrinology*, 15(4), pp.432-437. https://doi.org/10.1046/j.1365-2826.2003.00989.x

Carrillo-Vico, A., Guerrero, J.M., Lardone, P.J. and Reiter, R.J. (2005) A review of the multiple actions of melatonin on the immune system. *Endocrine*, 27, pp.189-200. https://doi.org/10.1385/ENDO:27:2:189

Chapoval, S., Dasgupta, P., Dorsey, N.J. and Keegan, A.D. (2010) Regulation of the T helper cell type 2 (Th2)/T regulatory cell (Treg) balance by IL-4 and STAT6. *Journal of leukocyte biology*, 87(6), pp.1011-1018. https://doi.org/10.1189/jlb.1209772

Cloyd, J. (2025) The estrogen-thyroid connection and its impact on women's health. *Rupa Health*. Available at: https://www.rupahealth.com/post/the-estrogen-thyroid-connection-and-its-impact-on-womens-health

Drucker, D.J. (2018) Mechanisms of action and therapeutic application of glucagon-like peptide-1. *Cell metabolism*, *27*(4), pp.740-756. https://doi.org/10.1016/j.cmet.2018.03.001

Fasano, A. (2012) Leaky gut and autoimmune diseases. *Clinical reviews in allergy & immunology*, 42, pp.71-78. https://doi.org/10.1007/s12016-011-8291-x

FDA (2020) *FDA review of GLP-1 receptor agonists for the potential risk of pancreatic cancer.* https://www.nejm.org/doi/pdf/10.1056/NEJMp1314078

Fröhlich, E., and Wahl, R. (2021) Physiological role and use of thyroid hormone metabolites—potential utility in COVID-19 patients. *Frontiers in Endocrinology*, 12, p.587518. https://doi.org/10.3389/fendo.2021.587518

Glatt, H. (2000) Sulfotransferases in the bioactivation of xenobiotics. *Chemico-biological interactions*, 129(1-2), pp.141-170. https://doi.org/10.1016/S0009-2797(00)00202-7

Gombart, A.F., Pierre, A. and Maggini, S. (2020) A review of micronutrients and the immune system–working in harmony to reduce the risk of infection. *Nutrients*, 12(1), p.236. https://doi.org/10.3390/nu12010236

Hagihara, Y., Yoshimatsu, Y., Mikami, Y., Takada, Y., Mizuno, S. and Kanai, T. (2019) Epigenetic regulation of T helper cells and intestinal pathogenicity. In *Seminars in Immunopathology* (Vol. 41, pp. 379-399). Springer Berlin Heidelberg. https://doi.org/10.1007/s00281-019-00732-9

Jochum, S., Moosmann, A., Lang, S., Hammerschmidt, W. and Zeidler, R. (2012) The EBV immunoevasins vIL-10 and BNLF2a protect newly infected B cells from immune recognition and elimination. *PLOS Pathogens*, 8(5), e1002704. https://doi.org/10.1371/journal.ppat.1002704

Kapatos, G. (2013) The neurobiology of tetrahydrobiopterin bi synthesis: A model for regulation of GTP cyclohydrolase I gene transcription within nigrostriatal dopamine neurons. *IUBMB Life*, 65(4), pp.323–333. https://doi.org/10.1002/iub.1140

Kelly, G. (2000) Peripheral metabolism of thyroid hormones: a review. *Alternative Medicine Review*, 5(4), pp.306-306. https://altmedrev.com/wp-content/uploads/2019/02/v5-4-306.pdf

Kerkvliet, N.I. (2002) Recent advances in understanding the mechanisms of TCDD immunotoxicity. *International immuno-pharmacology*, 2(2-3), pp.277-291. https://doi.org/10.1016/S1567-5769(01)00179-5

Lechner, J., von Baehr, V., Notter, F. and Schick, F. (2024) Osteoimmune Interaction and TH-1/TH-2 Ratio in Jawbone Marrow Defects: An Underestimated Association – Original Research. *Biologics*: Targets and Therapy, 18, pp.147–161. https://doi.org/10.2147/btt.s448587

Lechner, J. and Von Baehr, V. (2017) Silent Inflammation in the Jaw and Neurological Dysregulation-Case Study Linking Rantes/Ccl5 Overexpression in Jawbone with Chemokine Receptors in the Central Nervous System. *Journal of Dental and Oral Health*. https://scholar.google.co.za/scholar?as_q=&as_epq=Silent+Inflammation+in+the+Jaw+and+Neurological+Dysregulation-Case+Study+Linking+Rantes%2FCcl5 +Overexpression +in+ Jawbone+with +Chemokine+Receptors

Lee, H.C. and Wei, Y.H. (2005) Mitochondrial biogenesis and mitochondrial DNA maintenance of mammalian cells under oxidative stress. *The international journal of biochemistry & cell biology*, 37(4), pp.822-834. https://doi.org/10.1016/j.biocel.2004.09.010

Lieurance, J. (2022). *Melatonin: Miracle Molecule Beyond Sleep*. Primedia eLaunch LLC. https://books.google.mu/books/about/Melatonin_Miracle_Molecule_Beyond_Sleep.html?id=PDP3zgEACAAJ&redir_esc=y

Liu, K., Hong, B., Wang, S., Lou, F., You, Y., Hu, R., Shafqat, A., Fan, H.and Tong, Y. (2023) Pharmacological activity of cepharanthine. *Molecules*, 28(13), p.5019. https://doi.org/10.3390/molecules28135019

Lu, S.C. (2013) Glutathione synthesis. *Biochimica et Biophysica Acta (BBA)-General Subjects*, 1830(5), pp.3143-3153. https://doi.org/10.1016/j.bbagen.2012.09.008

Luyendyk, J.P., Schoenecker, J.G. and Flick, M.J. (2019) The multifaceted role of fibrinogen in tissue injury and inflammation. *Blood, The Journal of the American Society of Hematology*, 133(6), pp.511-520. https://doi.org/10.1182/blood-2018-07-818211

Mendel, R.R. and Schwarz, G. (2023) The history of animal and plant sulfite oxidase—a personal view. Molecules, 28(19), p.6998. https://doi.org/10.3390/molecules28196998

Moore, J. and Fung, J. (2016) *The complete guide to fasting: Heal your body through intermittent, alternate-day, and extended fasting*. Simon and Schuster. https://books.google.co.za/books?hl=en&lr=&id=8RBD-DQAAQBAJ&oi=fnd&pg=PA30&dq=%E2%80%A2%-09Fung,+J.+(2016)+The+Complete+Guide+to+Fasting:+Heal+Your+Body+Through+Intermittent,+Alternate-Day,+and+Extended+Fasting

Naviaux, R.K. (2014) Metabolic features of the cell danger response. *Mitochondrion, 16*, pp.7-17. https://doi.org/10.1016/j.mito.2013.08.006

Naviaux, R.K., Curtis, B., Li, K., Naviaux, J.C., Bright, A.T., Reiner, G.E., Westerfield, M., Goh, S., Alaynick, W.A., Wang, L. and Capparelli, E.V. (2017) Low-dose suramin in autism spectrum disorder: a small, phase I/II, randomized clinical trial. *Annals of Clinical and Translational Neurology*, *4*(7), pp.491-505. https://doi.org/10.1002/acn3.424

Nischwitz, D. (2020) *It's All in Your Mouth: Biological Dentistry and the Surprising Impact of Oral Health on Whole Body Wellness*. Chelsea Green Publishing.

https://books.google.co.za/books?hl=en&lr=&id=387S-DwAAQBAJ&oi=fnd&pg=PR9&dq=dominik+nischwitz

Pegg, A.E. (2009) Mammalian polyamine metabolism and function. *IUBMB Life*, 61(9), pp.880-894. https://doi.org/10.1002/iub.230

Pingitore, A., Nicolini, G., Kusmic, C., Iervasi, G., Grigolini, P. and Forini, F. (2016) Cardioprotection and thyroid hormones. *Heart Failure Reviews* 21(4), 391–399. https://doi.org/10.1007/s10741-016-9545-8

Pruessner, J.C., Hellhammer, D.H. and Kirschbaum, C. (1999) Burnout, perceived stress, and cortisol responses to awakening. *Psychosomatic medicine*, 61(2), pp.197-204. https://doi.org/10.1097/00006842-199903000-00012

Rayman, M.P. (2000) The importance of selenium to human health. *The Lancet*, 356(9225), pp.233-241. https://doi.org/10.1016/S0140-6736(00)02490-9

Riedler, J., Braun-Fahrländer, C., Eder, W., Schreuer, M., Waser, M., Maisch, S., Carr, D., Schierl, R., Nowak, D. and Von Mutius, E. (2001) Exposure to farming in early life and development of asthma and allergy: a cross-sectional survey. *The Lancet*, 358(9288), pp.1129-1133. https://www.thelancet.com/journals/lancet/article/PIIS0140-6736(01)06252-3/abstract

Rohleder, N. (2014) Stimulation of systemic low-grade inflammation by psychosocial stress. *Psychosomatic medicine* 76(3), pp.181-189. https://doi.org/10.1097/PSY.0000000000000049

Schmitt, EG and Williams, C.B. (2013) Generation and function of induced regulatory T cells. *Frontiers in immunology*, 4, p.152. https://doi.org/10.3389/fimmu.2013.00152

Schreiner, P., Harrer, T., Scheibenbogen, C., Lamer, S., Schlosser, A., Naviaux, R.K. and Prusty, B.K. (2020) Human herpes-virus-6 reactivation, mitochondrial fragmentation, and the coordination of antiviral and metabolic phenotypes in myalgic encephalomyelitis/chronic fatigue syndrome. *Immunohorizons*, 4(4), pp.201-215. https://doi.org/10.4049/immunohorizons.2000006

ScienceDirect (2005) Proteolysis. Agricultural and Biological Sciences. ScienceDirect. https://www.sciencedirect.com/topics/agricultural-and-biological-sciences/proteolysis

Segerstrom, S.C. and Miller, G.E. (2004) Psychological stress and the human immune system: a meta-analytic study of 30 years of inquiry. *Psychological bulletin*, 130(4), p.601. https://psycnet.apa.org/buy/2004-15935-004

Solerte, S.B., Fioravanti, M., Vignati, G., Giustina, A., Cravello, L. and Ferrari, E. (1999) Dehydroepiandrosterone sulfate enhances natural killer cell cytotoxicity in humans via locally generated immunoreactive insulin-like growth factor I. *The Journal of Clinical Endocrinology & Metabolism*, 84(9), pp.3260-3267. https://pubmed.ncbi.nlm.nih.gov/10487697/

Sompayrac, L.M. (2022) *How the immune system works*. John Wiley & Sons. https://books.google.co.za/books?hl=en&lr=&id=lVKMEAAAQBAJ&oi=fnd&pg=PR8&dq=Sompayrac,%E2%80%AFL.M.%E2%80%AF(2022)+How+the+Immune+System+Works.

Squizzato, A., Romualdi, E., Buller, H.R. and Gerdes, V.E.A. (2007) Thyroid dysfunction and effects on coagulation and fibrinolysis: a systematic review. *The Journal of Clinical Endocrinology & Metabolism*, 92(7), pp.2415-2420. https://doi.org/10.1210/jc.2007-0199

Stillwell, W. (2016) *An introduction to biological membranes: composition, structure and function*. Elsevier. https://books.google.co.za/books?hl=en&lr=&id=QW-pCwAAQBAJ&oi=fnd&pg=PP1&dq=Still-well,+W.+(2016).+%27Membrane+lipids+and+membrane+fluidity&ots

Straub, R.H., Cutolo, M., Buttgereit, F. and Pongratz, G. (2010) Energy regulation and neuroendocrine–immune control in chronic inflammatory diseases. *Journal of internal medicine*, 267(6), pp.543-560. https://doi.org/10.1111/j.1365-2796.2010.02218.x

Taneja, V. (2018) Sex hormones determine immune response. *Frontiers in immunology*, 9, p.1931. https://doi.org/10.3389/fimmu.2018.01931

Wiley, K.D. and Gupta, M. (2023) Vitamin B1 thiamine deficiency. *StatPearls Publishing*, Treasure Island (FL). PMID: 30725889. https://www.ncbi.nlm.nih.gov/books/NBK537204/

Won, E., Na, K.S. and Kim, Y.K. (2021) Associations between melatonin, neuroinflammation, and brain alterations in depression. *International journal of molecular sciences*, *23*(1), p.305. https://doi.org/10.3390/ijms23010305

Wrutniak-Cabello, C., Casas, F. and Cabello, G. (2001) Thyroid hormone action in mitochondria. *Journal of molecular endocrinology*, 26(1), pp.67-77. https://doi.org/10.1677/jme.0.0260067

Zhu, Y., Qiu, L., Jiang, F., Găman, M.A., Abudoraehem, O.S., Okunade, K.S. and Zhang, M. (2021) The effect of dehydroepiandrosterone (DHEA) supplementation on estradiol levels in women: a dose-response and meta-analysis of randomized clinical trials. *Steroids*, 173, p.108889. https://doi.org/10.1016/j.steroids.2021.108889

Zolkipli-Cunningham, Z., Naviaux, J.C., Nakayama, T., Hirsch, C.M., Monk, J.M., Li, K., Wang, L., Le, T.P., Meinardi, S., Blake, D.R. and Naviaux, R.K. (2021) Metabolic and behavioral features of acute hyperpurinergia and the maternal immune activation mouse model of autism spectrum disorder. *PloS one*, 16(3), p.e0248771. https://doi.org/10.1371/journal.pone.0248771

www.ingramcontent.com/pod-product-compliance
Lightning Source LLC
LaVergne TN
LVHW052304200926
848481LV00002B/459